DISCARD

SOLVING MATH PROBLEMS IN BASIC

No. 1564
$19.95

SOLVING MATH PROBLEMS IN BASIC

BY THOMAS P. DENCE

 TAB BOOKS Inc.
BLUE RIDGE SUMMIT, PA. 17214

Also by the author from TAB BOOKS Inc.

No. 1187   *The FORTRAN Cookbook*

FIRST EDITION

FIRST PRINTING

Library of Congress Cataloging in Publication Data

Dence, Thomas P.
    Solving math problems in BASIC.

    Includes index.
    1. Mathematics—Data processing.   2. Mathematics—
Problems, exercises, etc.   3. Basic (Computer program
language)   I. Title.   II. Title: Solving math problems
in B.A.S.I.C.
QA76.95.D46   1983   510'.28'542   83-4869
ISBN 0-8306-0264-X
ISBN 0-8306-0164-3 (pbk.)

Contents

Introduction

This book describes the interplay between computer programming and mathematics. It uses BASIC to help you solve problems that cover a very wide range of mathematics from number theory to algebra, analysis, probability and statistics. You are assumed to be somewhat familiar with the rudiments of programming; it doesn't have to be BASIC, although familiarity with BASIC or FORTRAN would be useful, and you should have some background in mathematics. This background should include at least three years of high school mathematics and preferably a quarter or semester of calculus.

This text is *not* designed to teach you how to program in BASIC. Some of the fundamentals of BASIC are discussed (variables, input, output, control statements, functions, arrays), and sample programs and flowcharts are occasionally given. For the most part, you will have to decide on a suitable approach to tackle the particular problem under discussion. I do not intend to lead you each step of the way. Occasionally this might happen, but not as a general rule. On the other hand, the book will expose you to a variety of mathematical problems suitable for programming, problems that are of serious intent and yet show promise for further work. You shouldn't read too lightly into this purpose. Exposure to a topic or problem does imply that some mental exercise is required of you. Not only will you have to construct the programs that solve the problems, but each section in the book contains an additional exercise, apart from programming, that is meant to test your knowledge and insight into the subject matter.

This book should fascinate those who have dual interests in programming and mathematics. The programming student needs more exposure to problems that involve algorithmic searches leading to something that may have a scientific application. The mathematics student, on the other hand, needs practice with programming because it is a valuable means for sharpening the thought processes. Needless to say, the combination of the two form a potent weapon for marketing yourself today.

The world is rapidly being engulfed in the computer revolution. Changes in our society are happening so quickly it is difficult to stay abreast. But there are significant and obvious reasons why this is happening. One, computers have so many important and useful functions that now the world could hardly do without them. And two, students of all ages, teenagers especially, have gotten into the act and have learned how to use computers, and how to build software. In the March, 1982, issue of *Money* magazine an amazing article, "Here Come the Microteens," reveals the tremendous impact teenagers are having as today's entrepreneurs in the software market. With this huge supply of technology potential, how can anyone predict the advances that await the world during the next two decades? What a pleasant problem. Furthermore, it is heartening to see some recognition on the part of our educational system toward these advances. Just last year, Rensselaer Polytechnic Institute gave each of 21 blue-chip (this term doesn't apply to just football players) freshmen (average S.A.T. scores almost 1500) his/her own $4,000 ATARI computer on which to do advanced work, or of course, play video games!!

Chapter 1
Computers

By now, practically everyone is aware of the tremendous impact computers have had on our society. Since the days of the Chinese abacus, an elementary counting device based upon a system of fives, man has continued to improve upon mechanical machines that could perform the basic arithmetical computations. When trading developed between different cultures, money was introduced into business relationships, systems for counting developed, and more complex economic situations evolved. As these events occurred, methods and tools were designed to aid people in accomplishing faster arithmetical calculations. Many such tools were developed during the seventeenth century by men like John Napier, William Oughtred, Blaise Pascal, and Gottfried von Leibniz. Over a hundred years elapsed before further significant advances were made. The Englishman Charles Babbage tried to build a steam-driven calculating device that would have contained all the major parts of a computer system as we know it today. His "Analytical Engine" was designed to be the first completely automatic general purpose digital computer.

Around 1855 a Swedish printer named George Scheutz built a machine similar to one Babbage had designed. This machine did perform elementary calculations, correct to 14 places, and was the first calculator to print out the results.

Near the end of the century, an American, Herman Hollerith, a statistician from New York, invented a machine that would help the government process data compiled for the 1890 Census. An electromechanical machine, the Hollerith Pantograph Punch, was used to reload, compile and tabulate digital data on about 63 million Americans.

The first electronic computers were developed in Great Britain and in the United States. The ENIAC (*E*lectronic *N*umerical *I*ntegrator *a*nd *C*alculator) was designed by John Mauchly and J. Presper Eckert in 1946 at the University of Pennsylvania. The machine contained an astonishing 18,000 radio vacuum tubes. Meanwhile, at Cambridge University in England, Maurice Wilks was building the EDSAC (*E*lectronic *D*elay *S*torage *A*utomatic *C*alculator). This was the first computer to use a set of operating instruction (programs) stored inside the computer. The mathematician John von Neumann [35] was responsible for this stored-program concept, certainly a fundamental concept.

Vacuum tubes were soon to be replaced by transistors. This enormous improvement in hardware was designed by a team of researchers at Bell Laboratories headed by William Shockley, Walter Brattain, and John Bardeen. This technology, called transistor circuitry, was not put to practical use until 1959. This initiated a second generation of computers because computing became so much more efficient. The transistor was 1/200th the size of the vacuum tube, used less than 1/100th of the power, and was more reliable.

Technological advances began to appear at a much faster pace. The second generation was soon to be replaced by a third generation. In 1964, just a scant five years later, the IBM System/360 was produced using integrated circuits (ICs). This complex circuitry was designed to solve a wider variety of problems. The IBM System 360 soon gave way to the compact minicomputers which, in turn, have been reduced to microcomputers. Microelectronics have become so advanced that an entire computer has been miniaturized to the point of being less than .25″ in size. These small computers are in vogue today, because of their cost, efficiency, and myriad of uses.

HARDWARE

Small computers soon could have memories that rival (if they don't already have) the capacity of the human brain. New electronic circuits give microprocessor rapid access to more than eight million bits, or basic units of information. Eight million is a very large number. On the average, the main memories of today's common microcomputer can store and recall only about 500,000 to a million bits.

The main memory of a computer consists of numerous electronic compartments, like mailboxes, each with its own address. The size of a computer memory, and consequently its usefulness for many important jobs, is limited by the number of digits in the box addresses. The microprocessors that do the work in such products as personal and small business computers can combine address digits in about 65,000 different ways. Any information that won't fit into that number of electronic boxes must be stored elsewhere (subordinate memory). Storing elsewhere, such as on recorders, typically means a cost savings, but at the expense of slower recall.

The major semiconductor makers plan to begin shipping microprocessors and related circuits that provide inexpensive computers with what is known as virtual memory. Devised by the British shortly after the EDSAC days and made popular by IBM during the 1970s, virtual memory is a complicated combination of programs and circuits that enables a computer to behave as if all the information at its command were stored in main memory when, in fact, most of it isn't. The virtual memory system calculates what information is likely to be needed next, moving it from disks into main memory and putting lower priority data back on the disks.

Battle lines are drawn all around the world as various companies strive to gain an edge on their rivals in the multibillion dollar computer war game. Today's electronic wizards have become highly proficient at cramming circuits onto sand derived silicon chips. One chip smaller than a baby's thumbprint can store as much information as a roomful of the vacuum tubes used in the early computers. The Japanese firm Fujitsu Ltd. has introduced a single chip 64K RAM (*R*andom-*A*ccess *M*emory), which stored 64,000 bits of data. This chip is on the way to becoming the standard for the new generation of computers, word processors, and office equipment.

It is also becoming the focus of the fiercest U.S.-Japan market battle yet. Although Japan has lagged in its prolonged push for dominant shares of the world wide computer and high-technology markets in general, its success so far with memory chips has been spectacular.

Japan's road to market dominance was helped by the 1973-75 recession, which retarded U.S. producers' output of 16K chips, and by Japan's then-growing reputation for making a superior, more reliable chip. Although that reputation was deserved in the past, analysts say, U.S. producers now seem to have closed the quality gap. But closing the market-share gap will be a bigger task.

The 64K RAM is expected to become the semiconductor industry's first billion dollar product in the next few years and to remain the industry's mainstay at least until 1986, when a 256K RAM is expected to be ready for the market.

Many agree that the Japanese companies enjoy certain strategic advantages in their drive for the memory chip market. Japanese companies have a lower cost of capital and a corporate mandate to focus on long-range product goals that frees them from quarter-to-quarter profit pressure. Moreover, Japanese universities are graduating large numbers of engineers at a time when America suffers an acute shortage of engineers.

Perhaps the biggest hurdle for the American manufacturers is overcoming the impression that the Japanese were producing semiconductors of superior quality. Three years ago, Hewlett Packard Co. publicly assailed the quality of American semiconductors purchased for use in its computers. Since then, American producers have solved the problem by instituting stringent quality standards and overhauling manufacturing procedures. But they are still fighting their old reputation. And they will have quite a fight trying to catch the Japanese firms of Fujitsu and Hitachi Ltd. The main competitors include Motorola, Texas Instruments, National Semiconductor, Intel, and Moster.

According to Erick Ayres, a marketing manager for Motorola, "We're going to beat the Japanese right where their strength lies—in the image of quality and reliability." Texas Instruments is even more confident. "We're sitting in the catbird seat," says Peyton Cole, the strategic-marketing manager for the memory devices. "We have a chance at being number one, at least by mid-1982," he says. But the sharp price cutting on the memory chips—from $20 in 1980, to $10 in 1981, and an estimated $6 to $8 now—makes it questionable whether there will be any winners in this market.

The bright spot for the American manufacturers is their technical skill. Even Fairchild Camera, a straggler in the 16K race, has designed an apparently outstanding 64K chip. "It's not going to be so easy for the Japanese this time," says Gilbert Amelio, Fairchild's general manager of microprocessor products [14]. "They may be surprised at the resolve some of us bring to this battle. It's going to be one hell of a fight."

The Japanese have long had a reputation as technological copycats: for taking a product (computer chips, automobiles) designed by someone else, making a few alterations to it, and then marketing it as something new. Now, to help erase this label, the Japanese have begun an intensive 10-year program to create new computers radically different from any existing today.

The Japanese call this the fifth generation of computers. The third generation of integrated circuits was mentioned previously. The fourth has come to be known as the very large scale integrated circuit, or VLSI. The Japanese expect the fifth generation to use, in addition to VLSIs, circuits that work on different physical principles from today's semiconductors. These powerful new circuits are "at the cutting edge of computer research in all aspects," says a U.S. computer expert. The research will encompass new hardware, new software, new architecture, and a new way of communications with human beings.

The Japanese machine would be able to hear and talk—at first in a special language and ultimately in natural human language. Unlike today's computers, which store and process data, the fifth generation computer will have "knowledge" with which to solve problems. Say the word *elephant* to it, and it will conjure up an image of a large, gray animal living in Africa, just as would the human brain. Tell it you'd like your payroll processed, and it will first write its own computer program to perform the task, and then do it.

One limit to the speed of computers is heat, the heat produced in the closely packed circuitry. The faster the machines add and subtract, the more likely they are to melt. But a new refrigerator may help keep them cool. The refrigerator, not much larger than the microelectronic circuits it is designed to chill, can produce temperatures as low as 310 degrees below zero, Fahrenheit. According to the inventor of the device, William Little, a professor of physics at Stanford University, improved versions may be able to approach absolute zero, the temperature at which some materials lose all their resistance to electricity and become superconducting. This would hasten the day of the superconducting computer without the need for handling such gases as helium or hydrogen that now must be used to keep the tiny circuits at the proper temperature.

The Japanese machine will probably have what is known as

5

"Josephson's Junctions," circuits that operate at temperatures almost equal to absolute zero. These circuits are much more rigid than the fastest semiconductors. Japanese scientists are developing these JJs as part of separate but related work on a supercomputer.

This research work is certainly not limited to the Japanese. Researchers at IBM's Watson Research Center in New York have been working on their own supercomputer for over ten years. The supercomputer under construction now is expected to process 10 to 12 times as much information per second as IBM's newly announced 3081 computer, their most powerful one ever. But even this supercomputer might only be a forerunner. Some IBM watchers believe that, by the next decade, the technology is capable of producing a computer with 50 times the computer power of the 3081, and the machine would still be smaller than a basketball. Incidentally, you are referred to [74] for an excellent account of the birth and growth of IBM, in particular, and the computer industry in general.

The critical step in the development of modern computers, is the cramming of about 1000 tiny electronic switches and circuits onto a chip about a quarter-inch square. Computers basically are nothing more than a vast collection of electrical switches that turn on and off to transmit or block an electrical current. A computer computes by turning the switches on and off in accordance with its program and the information fed into it. These switches are commonly transistor-like semiconductors made of such materials as silicon. They are tiny marvels that refuse to conduct an electrical current until it rises to a certain voltage. Then they go on, becoming conductors of the current. But no matter how fast the switches turn on and off, the electrical signals have to travel from switch to switch, and this takes time. To make computers faster—that is, more powerful—one needs to position the switches closer together. This is where semiconductors run into problems, because they emit heat. As the heat accumulates, the electronic components start to melt.

The ultimate aim of the IBM effort is a supercomputer that can carry out a whole computational step in a billionth of a second, a cycle time of one nanosecond. The key to this is a switch that is almost heatless. This is the same method that the Japanese are employing, for it is based on the concept known as Josephson's Junctions. This is named after Brian Josephson, an English physicist who won a Nobel prize for his work on superconductivity. A Josephson Junction is actually an electronic switch consisting of two tiny superconductors (IBM makes them of a lead alloy) separated by a thin film. When the current is low, it will flow from one superconductor to the next through the film, with perfect ease. If the current is increased or if a magnetic field is applied the thin film suddenly becomes an insulator, blocking the flow of the current. This way, the flow of an electrical signal can be turned on and off in a few trillionths of a second. Because they produce negligible heat, Josephson Junctions can be packed together extremely tightly.

There are several other areas where intense research is being conducted to improve the power of the computer. One effort is to replace the silicon in computer chips with a substitute, specifically gallium. Another is to alter the method of wiring the Josephson Junctions together, so that instead of producing the chips using the existing method of photolithography, scientists are experimenting with electron-beam lithography and x-ray lithography.

Relatively rare and expensive to extract, gallium is a bluish-white element obtained as a byproduct of aluminum and copper smelting. It isn't a metal, but it isn't a nonmetal either. It occupies that middle ground between conductivity and resistance which makes it a semiconductor, the raw material of integrated circuits. Gallium has been used since the 1960s in microwave transistors and since the 1970s in the illuminated digital read-outs on watches and calculators. The secret of gallium's superior speed lies in its atomic structure. Gallium has what chemists and electronic engineers call *high electron mobility*. At a given voltage, gallium's electrons move faster. The chips, commonly made of a compound of gallium and arsenic called gallium arsenide, offer circuit speeds two to five times greater than silicon. At highest speeds, gallium arsenide chips don't overheat and self-destruct as silicon chips do. However, gallium arsenide has its disadvantages. A three-inch wafer costs several hundred dollars, compared with $5 for a slice of silicon. It is quite rare, whereas silicon is more abundant than anything but oxygen. Gallium arsenide is also fragile and breaks easily. Some questions have even arisen about whether it might cause cancer.

The key element in making integrated circuits for random access memories and microprocessors is a printing process. Using photolithography, patterns of metal oxide are printed on silicon wafers in several layers. All together, the layered patterns form transistors, capacitors, and connectors. But this process of shining a light through a glass pattern mask has its limitations. Light waves can't focus on anything smaller than about three hundred-thousandths of an inch. This means that the lines and patterns of metal oxide on silicon for the next generation of microcomputers and memories are smaller than light can print. To cope with this problem, scientists have been experimenting with electron-beam lithography. These E-beam machines are a technological descendant of the electron microscope. Although electron-beam lithography makes for finer lines and patterns on computer chips, the machines are relatively slow and difficult to use. They are also quite expensive.

Electron-beam lithography is not the only way though of defeating the physical limits of light and breaking through the submicron barrier. Bell Labs and other industry researchers are working on X-ray lithography, where X-rays are substitutes for the conventional light source. In addition to their superior ability to make the precision lines, these X-ray machines are roughly 100 times as cost effective as photolithography. Some skeptics cite the inherent dangers of using X-rays too often as a drawback to this machine. Meanwhile other teams of researchers are working on a printing method using magnetically charged atoms, called ions and are claiming potential advantages over both electrons and X-rays. There is also a process called molecular-beam epitaxy, which dispenses with the printing process altogether. The aim of this process is to produce a sort of electronically active alloy in which atoms of several materials can be drawn up in alternating rows.

The Japanese, Bell Labs, and IBM are not the only folks pursuing construction of a supercomputer. The computer-science department at the University of Wisconsin at Madison has been awarded a National Science Foundation Grant to cover costs for building a supercomputer that would link the calculating and memory power of 50 to 100 computers. Even though this computer is of a different nature than the others mentioned, researchers say that such a system would represent a significant breakthrough in forming large networks of linked computers, which distribute the computing work over as many machines as needed. This idea is especially attractive to corporate computer users who would like to be able to talk to one another. Consequently more computer makers are designing machines to communicate with the big IBM computers that most large corporations already own.

In the Fall of 1981 several of the larger minicomputer makers, including Digital Equipment, Texas Instruments and Wang Labs, met and announced plans to support SNA, the systems network architecture that provide a framework for communication among IBM computers. This system allows minicomputer makers to line up with IBM's big machinery, thus providing service for both parties. Customers using IBM mainframe computers increasingly will have choices other than smaller IBM data processing machines for installation in plants and divisions. "Offering SNA is pretty important," says William Fachman, Director of Research for International Data Corp.. "Roughly 70 percent of the computers installed in large organizations are either IBM or IBM-compatible." William Davidson, a manager on Texas Instruments' digital systems group, says, "This gives us the opportunity to sell our products to companies that use an IBM." Some companies have experienced difficulty in developing the necessary programming to allow installation of SNA gear. Apparently it is no trivial task to make minicomputers that originally were designed to communicate in ways quite different from IBM's SNA-compatible.

Although such efforts are costly, the computer makers are charging relatively low prices. Honeywell, for example, is licensing SNA on its small computers for fees of about $300 a year. This is insignificant in contrast to the cost of the minicomputers themselves, which with a flock of terminals can run as high as $500,000. The minicomputers makers think they can absorb the costs of development because SNA compatibility will vastly expand their markets. Even personal or microcomputers are being linked by networks that provide users with much the same capabilities as the larger and considerably more expensive minicomputers. The personal computer market has recently blossomed into a multibillion dollar business.

Visionaries and computer manufacturers have long maintained that someday computers would be as common as telephones, and now that vision is becoming a reality. Powerful personal computers are on the threshold of taking over the office and transforming it irreversibly. Analysts estimate that of almost 500,000 personal computers sold during 1981, more than half were installed in offices.

The market for office automation gear, such as personal computers and word processors, is exploding. Until recently, Tandys' Radio Shack, Apple Computer, and Commodore International were among the leaders in the under $5000 computer market. But now hundreds of companies have entered the marketplace, from giants like Xerox, Hewlett-Packard, and IBM to those smaller companies currently selling copies like Monroe, Savin, Victor, Data General,

Pitney-Bowes, Wang, and Lanier. The list goes on and on. International Resource Development Inc., a Connecticut consulting and research firm, believes Japanese manufacturers will enter the U.S. home computer market in force with units able to use existing software programs and costing less than $300. IRD expects the Japanese to capture an incredible 40 to 50 percent of the small business market. These major Japanese firms include Nippon Electric, Sharp, Hitachi, Casio, and Toshiba. Furthermore, IRD estimates that only 5 percent of the potential market has been reached in the small-business and professional-technical sector, only about 2 percent of the market in the segment covering departments within corporations, and only about 1 percent of the vast potential home computer market. Even with all these potential markets, the competition may be so great that there will be minimal profit for anyone in hardware, while software grows strongly. IRD estimates annual software sales at $600 million in 1982, $2 billion in 1985, and $25 million by 1990.

The major manufacturers' goal is an all-electronic office where white collar productivity is sharply improved by a bevy of machines based on microprocessor technology that uses electronics to replace virtually every office function now done on paper. The array of services and equipment runs from electronic mail networks, electronic files, and word processors to laser printers and personal computers that can do sophisticated color-graphic design, financial analysis, or text editing. And the machines will be linked to communicate with each other. "The era of focusing on automating the secretary and clerical worker is drawing to a close, and the emphasis now is on automating white collar workers," says David Liddle, office-systems division vice president at Xerox.

A new piece of personal computer hardware was recently introduced by Tandy Corporation. Their new desk-top computer, called the TRS-80 Model 16, is much faster and more powerful than any of their existing models. Tandy's intent was to go after the professional, upper end of the personal computer market, which accounted for 90 percent of the $3.2 billion of computers sold in 1981. The base price for the Model 16 with a built-in eight-inch disk date-storage device and 128,000 characters, (which can be extended to 512,000) of main memory is $4,999. It is interesting to note that Tandy abandoned its traditional battleship gray color in favor of an ivory color for the Model 16. When you get into the more expensive computers, little things like appearance get to be almost as important as price and capability.

Now that the personal computer has become so popular, success may not be far away for the personal terminal. Many in the electronics and computer industries believe there is a large consumer market for small, portable machines with just enough electronic brainpower to talk over the telephone with large computers, enabling individuals to get information, do shopping and banking, pay bills, and play games. Much of the technology for such machines is available already. The question has been how to package it for the broadest, most profitable market.

During the summer of 1981 a small California company named Novation introduced a small battery-powered terminal, known as the Infone, that can fit into a briefcase and can telephone large computers almost by itself, sending and receiving messages even when its owner isn't around or late at night when telephone rates are lower. Touch typing is possible on the Infone, which can store about three pages of text internally and much more if it is attached to a tape recorder. A telephone handset about the size of a fountain pen is part of the machine, which also contains circuits that produce electronic speech. It can send and receive voice as well as text and can make recorded announcements. The machine also can serve as a telephone dialer and can turn household or office appliances on or off at preselected times. The first models of the Infone were able to display only one line of text at a time on a panel of liquid crystals, but a printer can be attached. The company planned on adding a connector that would allow the use of television screens. Initial prices of the Infone averaged about $1000, but the costs were expected to drop considerably by this year.

Tiny terminals have obvious disadvantages. Typically, the smaller they become, the less information they can retain or display and the more difficult their keyboards are to use. The California firm of MSI Data, with years of experience in making small terminals for industrial and commercial markets, doesn't believe a mass-market for such terminals exists. This belief was partially borne out last summer when Nixdorf Computer tried to turn its hand-held electronic language translator into a hand-held terminal, but stopped selling the product after marketing it only a few months. This pocket translator was quite a nifty device. Selling for as low as $49.50, the LK-3000 was capable of translating as many as 11 different languages, performing metric conversions, computing foreign exchange, organizing filing systems, maintaining notes in its memory bank, and even supplying all the facts regarding past summer and winter Olympics.

As a firm or an individual becomes more dependent on computers, problems can arise when computers break down. One way that the risk can be averted, but only at great expense, is to employ two computers where one acts as a back-up that can go into action if the first fails. Seeing a market opening, several firms have recently introduced computer systems that are billed as practically fail-safe. Tandem Computers introduced the first such system, its Non Stop system, in 1976. The company says Non Stop has become a phenomenal success. Pitney-Bowes Corp., the postage-meter company in Connecticut, started using a Tandem system in 1979. Pitney-Bowes was marketing a remote postage-meter resetting system by which a computer could increase the amount of postage available to a user by telephone line. The service required approval of the U.S. Postal Service, which demanded reliability. Customers also wanted assurances of reliability. Pitney-Bowes says the Tandem it installed worked accurately "over 99.9% of the time." Two other firms that have followed Tandem into the fail-safe market include Stratus Computer Inc. and Sequoia Systems Company.

Studying the makeup and composition of a computer system is not something that the typical person is interested in. He or she, instead, is interested in how the computer will affect his or her life and what computer applications they need to be concerned with. Let's take a closer look at some recent developments.

APPLICATIONS

Computer applications are as unique as each computing machine, its software, the user, and the problem to be solved. The technical limitations of computing systems are challenged daily by the imagination of each user. Computer technologies and creative thinking have worked together to solve millions of problems. Computers can not only add, subtract, multiply, divide, and compute logarithms and exponentials, but they also help to sort, weigh, move, and label enormous quantities of things. Computers help people to position aircraft, to find lost cities, and to saw logs. They are used to help compose music, to monitor volcanoes, to weld metal, and to design bridges. Computers are being used more and more to improve the ways in which America's needs are met.

To begin with, there could very well be a significant change in the house of the future. Many new innovations that stem principally from new technology, changing economics, and different lifestyles are planned. One change appears obvious—increasing use of solar design. Other changes center around the materials to be used in specific constructions. Homes are expected to become smaller, thus necessitating more efficient use of each room. This is where the computer offers its help. It is quite likely that many plumbing, heating, and lighting systems will be under total electronic control. Some computers will thus monitor the home electricity, plumbing, heating, cooling, and lighting. Based on information fed into the system, it will be able to adjust automatically to the needs of the occupants and run by itself without even a finger on the thermostat or light switch. A device can, for example, read the temperature and humidity: if the weather gets too hot, it can open windows, turn on an evaporative cooler, or start the air conditioner. Further improvements may see the end of the house key; a resident punches a number code into the computer at the door. The computer may be programmed to welcome family members by name. Sensors in the house can turn lights off and on when somebody enters or leaves the room. They can also detect an intruder. The possibilities are almost limitless, and will certainly make for an interesting era in interior design.

The Federal Government is as interested in technological advances as anyone, especially if it concerns the military. For instance, the Chrysler Corporation has been working on the MI tank for the past ten years (in addition to which they still produce the M60 tank, now 21 years old). Even though Chrysler has been besieged by many financial problems, this hasn't altered the technology built into the MI. Virtually everyone agrees that if the MI can be built in quantity to specifications, it will be the best tank in the world. In addition to increased speed and power, the MI's fire-control system enables it to routinely hit targets more than a mile away while bounding over an obstacle course like a dune buggy, an unusual capability for a tank. The highlight of this feature is that the gunner uses a laser range finder to line up the target. Then a computer makes automatic adjustments, and the gunner need only squeeze the trigger. This increases the accuracy over the M60 by almost two-fold.

The Federal Government has also stepped up its budget allowance to the Defense Department, the Federal Aviation Administration, and in particular, the National Weather Service for improvements in their radar equipment. The National Weather Service had an elaborate $340 million plan to replace its existing network of conventional weather radar with new machines, called Doppler machines, by the end of this decade. The old radar equipment was not only wearing out but was technologically outdated. The new radar takes advantage of a phenomenon first described by the Austrian physicist Christian Doppler in connection with light waves and sound waves. Radio waves coming from an object moving toward you have a higher frequency than those coming from an object at rest; radio waves coming from an object moving away have a lower frequency. A Doppler radar sends out its radio beam at a rigidly fixed frequency, unlike the sloppier signal of a conventional radar. As windswirled raindrops bounce the signal back, receivers at the radar station detect the higher frequencies for those twisting inward and lower frequencies for those twisting away. A computer is used to translate the contrasting frequencies into different colors that make a distinctive pattern on TV monitors, allowing human operators to see a tornado being born in clouds up to 140 miles away.

The new radar network is planned for improved forecasting of several kinds of bad weather, not just tornadoes. According to Arthur Hansen, the National Weather Service official in Silver Spring, Maryland who is directing the change over, "the biggest payoff is in warnings of flash floods, which cause more deaths and property damage than tornadoes." The flash flood predictions will have nothing to do with the Doppler effect. Rather, the government is telling potential contractors for the new radars to design computers for them that can estimate the amount of rain falling in a severe storm and match that estimate with the shape of local watersheds. The specifications also tell bidders to come up with computer programs that will improve forecasts of airplane-threatening wind gusts that precede thunderstorms, and of storms likely to generate heavy hail.

Computers are being used more and more in our search for additional fuel reserves. Vast resources of both oil and gas lie undiscovered beneath the earth's surface. To help search them out, Exxon has engineered a system that uses the eye of a satellite. The satellite, one of NASA's Landsats, circles the globe every 103 minutes, scanning the landscape below. Its eye measures the intensity of reflected sunlight acre-by-acre and radios the information back to earth. Giant computers at Exxon convert this information into pictures of the earth's surface. The pictures are then color-coded by the computer to reveal details that could never be seen by a human eye. These pictures help Exxon geologists locate the areas most likely to hold deposits of oil or gas.

Many functions have become computer-assisted for the first time. It might prove interesting to you to hear of just a few of them. For one, last summer witnessed a computerized Tickertron campground reservation service initiated by the National Park Services. National parks on the reservation system include the Grand Canyon, Great Smokey Mountains, Rocky Mountains, Sequoia, Shenandoah, and Yosemite. More parks are likely to follow if motorists continue to visit these national gems in increasing numbers.

Futhermore, the Supreme Court has also advanced into the age of computerization. In October of 1981, for the first time since the Justices began issuing opinions in 1972, decisions began being printed by a computer. The changes were supported by Chief Justice Warren Burger, who recalls that when he arrived at the Supreme Court in 1969, "They didn't even have a Xerox machine in the building."

Can you imagine the Old Testament Book of Genesis being analyzed by a computer? During the winter of 1981, results were revealed of a computer analysis of the book's 30,000 words. Results, which indicated, according to Yehuda Radday, professor emeritus of Biblical studies at the Technion, Israel's Institute of Technology at Haife, that there was an 82 percent probability that Genesis was written by a single author. A four man team led by Dr. Radday and made up of a professor of computer science, a statistician, and a mathematician, thoroughly investigated word use and word frequency in the book. They believe the computer investigation revealed the stylistic fingerprints of only a single author. On the other hand, the question of the infallability of the computer was raised by another biblical scholar who felt that certain nonlinguistic differences in style might not have been picked up by the computer.

Surely everyone is familiar with some of the new innovations in automotive technology, such as a computerized system that measures the efficiency of the fuel system, or the electrical system. Now there is a computer car starter on the market (for about $330) that allows you to start your engine by remote control, which is helpful in cold weather. Just push a button from up to 500 feet away and the unit will turn on the ignition, pump the accelerator, and shut off the engine when it is warm.

Computers have made their entry into the sports world in several ways. The National Collegiate Athletic Association (NCAA) began using computers to help select teams to play in its Division I basketball championship. The computer ranked each of the 264 Division I teams according to four factors: winning record, strength of the opponents' schedules, opponents' success, and success in games played on the road. Each of these factors were assigned a particular weighting, and the computer compiled weekly data and computed the rankings accordingly. In the sport of golf, many of the leading equipment manufacturers are using extensive computer machinery to test and analyze their new hardware. But it wasn't until last year at the Tournament Players Championship at Sawgrass that the PGA officials unveiled their new computerized scoreboard system. The Vantage Scoreboard, sponsored by the Reynolds Tobacco Company, featured a large main board alongside the 18th green, and 29 smaller ones properly placed on the course. The boards would supply the spectators and golfers with all the necessary information about the scores of the leaders. The scoreboards could also display tournament history and attendance figures, and reproduce photographs, logos, and other graphics much as the electronic scoreboards in baseball and football.

When April rolls around, many people in this northern portion of Ohio begin avid preparations for spring housecleaning, yardwork, and gardening. To many, though, it signifies the last few weeks before the income tax returns are due. Microcomputer programs to handle these tax-filing chores have been on the market for several years. But until recently, most of them were too complicated to be practical for most home-computer owners. Designed primarily for accountants and lawyers, they typically included a lot of arcane schedules and forms that few tax payers regularly need. Expecting widespread interest in Reagan's new tax law, however, software makers this year have come up with streamlined versions of these programs aimed at almost any taxpayer experienced at filling out his/her own return. The programs can do everything from the Short Form 1040a for individuals who don't itemize to income averaging and reporting complicated business transactions or tax-shelter investments. One such tax-filing program is the *Tax Prepayer* that will do your state income tax return as well as your itemized Federal 1040 and all of its supporting attachments. The package is $150 and is sold through Howard Software Services of La Jolla, California. Other packages include the *Tax/Saver* by Micromatic Programming Company, which sells for $120, and Panasonics' *Tax Model*, which retails for $149.

Did you ever wonder what happened to the dinosaurs? Many theories have been proposed suggesting plausible factors leading to their demise. A couple of researchers (Thomas Ahrens and John O'Keefe) at the California Institute of Technology recently presented their findings on this subject. According to their computer model, a large comet, or meteor, say seven miles in diameter, struck the earth and landed in the ocean. This caused a three-mile high tidal wave that flooded many areas of the earth, in particular the marshy areas that dinosaurs frequented. Upon impact, material from the meteor would have been thrown into the air, forming a huge dark cloud to circle the globe. This prevented much sunlight from reaching the surface, which disrupted the entire food chain of plants and animals. The presence of the rare element iridium found at various places around the earth tends to support these findings. The reader is referred to the April 7, 1982 issue of the Chronicle of Higher Education for further details on this interesting proposal.

One of the more amazing campaigns today is the work that is being done to direct computers to duplicate both human speech and hearing. Scientists have been studying how machines can be taught to talk and respond to voice commands. Research conducted at Bell Labs on the human larnyx and vocal tract has shown scientists how to program computers to talk or synthesize voices. These voices can speak electronically stored information, such as telephone numbers, calendar dates, inventory data, and airline timetables. And the computer can be taught to listen so that, for instance, it is able to verify who is talking to the machine.

One unique byproduct of the computer age is the slow but growing change in the work habits of people. Specifically, with the advent of the personal or microcomputer, more people are able to conduct a sizeable portion of business from their own home. A home computer can be hooked up with telephone equipment to huge data banks and allow the user to contact almost anyone and anything. People can engage in programming, technical editing, or writing with the proper equipment. Setting up shop at home, otherwise known as the electronic cottage (a term coined by Alvin Toffler in his book "The Third Wave"), has some distinct advantages: there is no traffic problem getting to work and no special wardrobe to wear, and the Internal Revenue Service offers tax benefits for the home-based businesses.

With all these advances in computer know-how, it is not surprising that all the educational systems around the country have modified their curriculums to meet these new demands. Schools are purchasing hardware so they can offer technical training to their students. The idea that all students "should be acquainted with the computer in some reasonably respectable fashion is surely no more radical a thought than the proposition that they should be able to read and write," says Stephen White, director of special projects for the Alfred Sloan Foundation.

At several college and university institutions, "computer literacy" is now required for graduation. Computer literacy will be a requirement for all students at Hamline University, starting with the freshman class in 1982. Rochester Institute of Technology has adopted a set of goals for computer literacy for both students and faculty members. After R.I.T. made this announcement, some 300 full-time faculty members applied for a two-week summer workshop to help them understand and appreciate computing.

These requirements seem reasonable when you stop to tally up all the functions a computer system has control of. Less than three years ago, for instance, Bowling Green State University relied on just one computer on campus (with an enrollment of about 18,000) to meet all its academic and instructional needs. Now, however, computers and terminals have been added in nearly every area of the campus so that they do everything from registering students and printing paychecks (and depositing them in banks) to monitoring building temperatures.

One of the large computer firms, Control Data Corporation, has been highly effective in providing educational assistance for many disadvantaged people. Control Data has spent considerable time and money in building an effective computer-based education. Their prominent PLATO system proved especially useful for raising the academic standards of Los Angeles' Berendo Junior High School. The school now has a showcase of city, state, and national awards for educational achievement. The PLATO system has also proved highly effective in penal and correctional institutions where inmates are typically a distance apart from some form of academic training. The PLATO system allows inmates to take themselves through special lessons by direct interaction with a computer. It monitors their answers, responds to their needs and questions, and compliments their progress or criticizes an incorrect decision. Control Data has a complete curriculum available for marketing to penal and correctional institutions in the country. Furthermore, Control Data developed their HOMEWORK program some four years ago to help train and provide jobs for disabled people. Many people confined to their homes because of a physical handicap now have the opportunity to carry on productive work because of the training supplied at the HOMEWORK project.

One revolutionary idea in education is in the change of format of encyclopedias. Instead of the burdensome 26 volume set of hardback bound books, soon you will have the chance to purchase the computer tape that contains all the standard information. These tapes can then be hooked to your home television and the information displayed on the screen. Both World Book and Britannica have been exploring and researching these new-frontiers for the past several years. The most active experimenter, though, is a newcomer, Arete Publishing Company, which in addition to its 21-volume Academic American Encyclopedia, introduced in August 1980, has all the same information on computer tapes. These tapes not only display the information concerning items such as bird calls, music pieces, and human speech, they can produce the appropriate sounds. These electronic encyclopedias have other advantages over printed ones, in addition to sound. The most important is that they can be updated almost instantly. Just think how beneficial it would be for a student to watch an hour of "encyclopedia television" each night instead of Laverne and Shirley.

Reference companies are looking beyond encyclopedias. Source Telecomputing Corporation in Virginia wants to put the Bible into the computer data base of its 8000-subscriber information network. "A lot of people would like to summon their favorite biblical passage onto their home computer terminal," says Bettie Stieger, Source's vice president of information resources.

Apart from the Japanese, many other foreign countries have been developing their computer resources to help meet the growing needs of society. France recently revealed details of a multimillion dollar world center to design computer systems for education and training in industrialized countries and in the third world. This new World Center for Microcomputer Science and Human Resources is to reflect the Mitterand government's strong belief in the computer as an agent for social change while further extending efforts to compete against the United States and Japan in the burgeoning world computer telecommunications market. The center is expected to place a heavy emphasis on computer education and the third world. Pilot projects are tentatively scheduled for Senegal, Kuwait, Ghana, and the Phillipines.

Computer growth in the last decade has been phenomenal, and there is no reason to expect any let-up during this decade. With all the people and students that are currently exposed to some form of computer use, the growth will most certainly continue. According to a survey conducted by Market Data Retrieval of Westport, Connecticut, there were microcomputers in some 22,000 public schools and school districts as of Fall, 1981. Senior high schools possessed the majority, but surprisingly enough, almost as many elementary schools have computers too.

Unfortunately not all the news concerning computer growth is good news. As with any technological advancement, there are those who find new methods and more ingenious methods for applying the advancement to their own personal gain at the sacrifice of others. We are witnessing the dawn of the computer crime era, an era in which it has become possible to steal large sums from a company without laying hands on the money itself. The computer can make embezzlement easier to accomplish and harder to detect. With handwritten records, usually only trusted employees have the opportunity to embezzle because only they have access to the books. But computerization means all the records an embezzler needs to manipulate are accessible at one place, the computer console. Thus, the clerk hired to make routine entries at the console could be stealing the place blind. Computer cheats alter accounts payable and receivable to divert money for themselves or juggle payroll data to get checks for phantom workers. This electronic fraud holds several advantages for the thief: one, he usually has access to larger quantities of funds than before; two, computer fraud is easier to hide; and three, if caught and convicted, the penalty is no more severe than before.

One of the latest sources of computer crime is software piracy. Producing software, commands on magnetic tape or floppy disks telling a computer what to do, was a $3 billion business just last year. By 1985, sales of programs that make video spaceships eat asteroids, instruct computers to process a paycheck, or plot a financial forecast are expected to hit $15 billion. And the opportunity and temptations to copy the programs instead of buying them are on the rise, too. Piracy is easy, cheap, and quick—just pop a software diskette into a computer in the office or at home and tell the machine to copy.

Donn Parker, a senior managements systems consultant for SRI International, says software piracy is an even bigger problem than the theft of the tiny semiconductors, known as silicon chips, that amounted to losses of about $2 million last year. And more sophisticated than this are the video game imitations. "Millions and millions of dollars have been lost nationwide . . . and nearly every coin-operated game has been copied, pirated, or modified," says Paul Laveronia, a lawyer who represents ATARI, the largest maker of computer games. At least these video game imitations don't present any problem of national security, whereas thefts of the chips might. This type of thievery has been, and still might be, a common occurrence in many of the scientific corporations located in the technology-rich "Silicon Valley" of California. Just last spring someone apparently walked in off the street and stole $64,000 worth of semiconductor chips from Advanced Micro Devices, Inc. The underlying problem here was that the stolen circuits had been designed to military specifications for use in the nations' missile and aerospace program. The fear is that the chips will turn up in somebody else's missiles—more than likely the Russians.

Thefts and diversion of electronic components have occurred at such major producers as Intel, Texas Instruments, and National Semiconductor, as well as to many smaller companies. The cost of thefts to Silicon Valley companies alone has been averaging $20 million a year. That there is a hunger for these hi-tech goods behind the Iron Curtain is unquestioned. Although the Soviet Union is well ahead of the U.S. in conventional military firepower, it is believed to lag behind in its electronic arsenal. Thus, the U.S. Government has made it illegal to export certain high-technology equipment to the Soviet Bloc and other unfriendly nations without a Commerce Department License.

The topic of national security has touched off some violent debates of late. In particular, those scientists and researchers that have studied and worked on problems involving computer security are being forced (asked??) to share their information with the government because, if published, their results may prove useful to unfriendly powers. Several scientists have already developed intricate codes that could hurt national security if the studies were published. A foreign power may use the findings, for instance, to read secret U.S. messages. On the other hand, the research could give a foreign government a better coding system than it is now using, making it harder, perhaps impossible, for the National Security Agency to decipher that government codes.

Another threat to national security comes in the form of electromagnetic pulses, or EMP. These are intense waves of energy released when, say, an atomic bomb explodes high over the earth. These waves are powerful enough to disrupt a country's electrical power and computer network. Nuclear weapons experts have known about EMP for decades, and steps have long since been taken to protect military equipment. But most civilian systems are unguarded. The threat to national security has increased in recent years because of the increased power of nuclear weapons and because both military and civilian equipment increasingly have come to depend on microelectronic circuits, which are far more susceptible to EMP than such other technologies as vacuum tubes and electromechanical relays. Clearly, a powerful EMP attack might cause unprecedented damage to the country.

Chapter 2
BASIC

BASIC (Beginner's All-purpose Symbolic Instruction Code) is a computer language developed at Dartmouth College under the direction of Professors John Kemeny and Thomas Kurtz. This is the language that will be used for all the mathematics that follows. BASIC is well suited to handle many of the mathematical operations and iterative schemes that are present in most undergraduate mathematics and some graduate mathematics. Because the BASIC language is composed of easily understood statements and commands, it is one of the simplest programming languages to learn. Along with FORTRAN, it provides enough structure to be highly suited for many of the mathematical needs. It is important to keep in mind that there are many mathematical calculations that no computer can do. This becomes obvious when one contemplates all the abstractions involved in mathematics. You may be attempting to show that a particular vector space is actually an inner product space, or you may be trying to develop a topology on a space, or you may just be trying to evaluate the limit.

$$\lim_{n \to \infty} \sum_{i=0}^{n} \frac{2^{2i+1}}{(2i+1)!} \quad .$$

Computers cannot handle operations that involve the infinite (e.g., the above limit, or computing a derivative), but they are adept at performing an arbitrarily large number of arithmetical computations. This proves very useful in many phases of mathematics where an approximate value is sufficient. All of the mathematics problems to be discussed from here on out will lend themselves to this treatment. Before the mathematics are discussed, some of the statements and features of BASIC will be described. By no means will this be an exhaustive list, for nothing will be mentioned concerning string variables, conversion functions, or files. The features to be discussed can be classified according to the following five headings:

1. constants, variables
2. input, output
3. control statements
4. functions
5. arrays

It is most important to mention that there are many different versions of BASIC. They all have a majority of language commands and terminology in common, and they all have a few differences (usually just minor differences, such as denoting the natural logarithm function by LOG or LN). I worked with the BASIC-PLUS-2 version implemented by Digital Equipment Corporation, and hence, what follows, is their particular system.

There are three types of constants in BASIC, namely numeric, integer, and string. The numeric constants, also known as the real numbers or floating point numbers, consist of one or more decimal digits, either positive or negative, in which the placement of the decimal point is optional. The following are all valid numeric constants:

51	−1234	13.
7.4	.80	13

BASIC only accepts numeric constants that fall within a certain range. Any number outside the range will produce an error message to that effect. However, you can input very large and very small numbers by resorting to scientific notation. Here, use the format

$$x.xxxxx \text{ E } (+ \text{ or } -) \text{ } n$$

and, for example, you could write

$$2.31472 \text{ E } 13$$

to represent the number 2.31472×10^{13}. You could also write 1.12345 E-7 to represent the small number .000000112345.

Integer constants are whole numbers written without a decimal point, and followed with a percent sign, %. For example, the following numbers are all integer constants:

31%	−12%
7%	54321%

The following are not integer constants:

3.1%	−.12%
7	5.44444E8.

And again, integer constants must fall within a specified range, or BASIC will reject them.

BASIC accepts four kinds of variables, namely numeric, integer, string, and subscripted. A numeric variable is a named location in which a single numeric value is stored. You name a numeric variable with a single letter followed by 29 optional characters consisting of letters, digits, or periods (no blanks!). The following are valid numeric variables:

A	B1H42
A3	ZZ..3
HOUSE	TOM

Typically one uses combinations of letters and digits so as to give meaning to the quantity in question. It could make more sense to the programmer to refer to, say, a boy's age by using the variable AGEBOY, rather than using R2D2.3H. There are some key combinations of letters that you are not allowed to use as variable names, because these words have already been built into the system with another meaning. Some of these key words include:

ACCESS
BIT
DATA
DELETE
END
ELSE
ERROR
FILE
FIX
FOR
GOTO
IF
INV
LET
LOG
MOD
NOT
PI
RANDOM
RETURN
SGN
SIN
SQRT
THEN
TAN
UNTIL
VAL
WAIT
WHILE
ZER

Before program execution, BASIC assigns the value zero to all numeric variables in your program, except those you declare otherwise. An integer variable is represented by the same set of characters as a numeric variable, but it must be followed with a percent sign, %. Thus A% and HOUSE% are integer variables, while A, 2HOUSE% and B2 are not. Integer variables always represent the whole number portion of whatever number is assigned to them. Thus, the statements

$$A\% = 2.8$$
$$B\% = -2.8$$

would be machine interpreted as assigning the whole number 2 to A% and the number −2 to B%. You should note that this process is not equivalent to rounding off to the nearest integer or to the greatest integer function studied in mathematics.

A subscripted variable is a numeric, integer, or string variable with a maximum of two subscripts appended to it. (This is a serious drawback in many cases, and represents a significant difference from FORTRAN.) To name a subscripted variable, you just use a numeric, integer or string name; but when referring to a particular element in the list, parentheses must follow. For example, A(4) refers to the fifth item held in storage by the numeric variable A; A(4) = 37. When two

A(0)	A(1)	A(2)	A(3)	A(4)
3	7	15	21	37

subscripts are used, like A(2,3) the first subscript designates the row number, and the second subscript denotes the column number. If a subscripted variable is used in a program, a DIM statement is required to define the dimensions of the variable.

You can form arithmetic expressions using these constants and variables because BASIC allows for the fundamental operations of addition, subtraction, multiplication, division, and exponentiation. If A and B are two sample numeric variables, the following expressions can be formed:

Expression	*Meaning*
A+B	Add B to A
A−B	Subtract B from A
A∗B	Multiply A by B
A/B	Divide A by B
A^B	Raise A to the power B
A∗∗B	Raise A to the power B

If several operations are combined in a single expression, BASIC performs those enclosed in the innermost parentheses first and then moves to the outer parentheses, just as you handle typical algebraic expressions. If several operations are enclosed in the same parentheses, BASIC evaluates the expression according to a hierarchy of operations. In this hierarchy, exponentiation draws the top priority, followed by multiplication/division, and then addition/subtraction. Thus, the constant

$$2+3\wedge2*4$$

is evaluated

$$2+3\wedge2*4 = 2+9*4$$
$$= 2+36$$
$$= 38.$$

There are several methods of supplying data to a program in addition to the obvious way of listing each variable with its associated value.

$$10 \ A = 2.1$$
$$20 \ B = 3.7$$
$$30 \ C = \ \ 41$$

The two standard methods involve the input statement, and the read/data statements. The input statement has the format

INPUT variable(s)

where each variable is separated by a comma. When you run your program, BASIC stops at the line designated by the input statement and prints a space, a question mark (?), and a space. BASIC then waits for you to type one value for each variable requested in the input statement. The above three variables and their values could be fed into the computer as follows:

```
10   INPUT A,B,C
.
.
.
100   END
READY
RUNNH
?2.1,3.7,41
```

The input statement is advantageous because it allows you to easily change the values assigned to the variables listed in the input statement. On the other hand, it takes a little longer to run the program because the input statement requires that you interact with the computer while the program is running. This does not happen

28

with the read/data statements. In this situation you supply a pool of data to the program in advance, thus building a data block. You must first supply a read statement, which lists all the variables in question (any combination of numeric, integer, string, or subscripted variable is allowed: for example, 10   READ A,B,C%. The computer is then directed to the data statements, found anywhere later in the program, which list the values of these variables. Thus, you might have the line, 90   DATA 2.1, 3.7, 41. The computer will deliver an error message if you omit the data statement or if you do not list at least as many values in the data statement as there are variables listed in the read statement.

When it comes time to produce output, the print statement comes into play. There is some versatility with this statement that allows you a degree of flexibility in the form and style of the output. For instance, typing **PRINT** followed by a variable causes the computer to print out the value of the variable. For example, the program

```
10   A = 2.1
20   PRINT A
30   END
```

would print the output

```
READY
RUNNH
2.1
```

Typing **PRINT**, followed by the variable in quotation marks causes the computer to print whatever characters are enclosed by the quotation marks. Thus we could have

```
10   A = 2.1
20   PRINT "A"
30   END
READY
RUNNH
A
```

Typing the word **PRINT** followed by nothing causes the computer to leave a blank line. For example,

29

```
10   A = 2.1
20   PRINT "A"
30   PRINT
40   PRINT A
50   END
READY
A

2.1
```

You can print the values of several variables on the same line by typing **PRINT** followed by the variables, where the variables are separated by commas or semicolons. The comma causes the items to be spaced apart, using printing zones (each zone contains 14 spaces), while the semicolon causes tight packing on the print line. For example:

```
10   A = 2.1 \ B = 3.7
20   PRINT A, B
30   PRINT A; B
40   END
READY
RUNNH
2.1                  2.7
2.1    3.7
```

These are other methods that can be used to help position the output and control its appearance. One method makes use of the tab function in conjunction with the print statement, and another is a print using statement. The reader is referred to a manual for clarification on each.

Control statements allow the programmer to change the sequence of execution steps, usually depending on what values specific variables assume. The GOTO statement (which could also be written GO TO) causes control to be immediately transferred to the statement that it identifies. For example:

```
10   A = 2.1
20   GO TO 40
30   A = 17.3
40   PRINT A
```

```
50  END
READY
RUNNH
2.1
```

We note here that line 30 was never executed because line 20 directed us to skip directly to line 40.

The if-then statement provides for a transfer of control depending on the truth of a conditional expression. The format of this statement is

IF (Conditional Expression) THEN (line number).

If the conditional expression is true, the computer transfers to the line number so specified after THEN; if the expression is not true the computer transfers to the next line in the program. This can be illustrated with the following:

```
10  A = 2.1 \ B = 3.7
20  IF A<3 THEN 40
30  PRINT "HELLO"
40  PRINT "HI MOM"
50  IF B<3 THEN 70
60  PRINT "GOODBYE"
70  END
READY
RUNNH
HI MOM
GOODBYE
```

Line 30 was skipped because A was less than 3 and line 50 was ignored since the value of B was not less than 3.

A third control statement is the for-next statement, which is extremely important because it allows for repeated execution of a set of statements forming a loop. The loop saves you from unnecessary duplication of steps. Suppose you wished to add the odd numbers 1 through 19.

$$SUM = 1 + 3 + 5 + 7 + 9 + 11 + 13 + 15 + 17 + 19.$$

The for and next statements can be used to evaluate this sum as follows:

```
10   SUM = 0
20   FOR I = 1 to 19 STEP 2
30   SUM = SUM + I
40   NEXT I
50   END
```

The program causes I to first assume the value 1. SUM will be computed as SUM = 0 + 1 = 1. Then I will increase by a step of 2, so I = 3. Then SUM will be recomputed as SUM = 1 + 3 = 4. Then I becomes 5 and SUM = 4 + 5 = 9. Continuing, SUM = 16, SUM = 25, SUM = 36, SUM = 49, SUM = 64, SUM = 81 and finally when I = 19, SUM = 100. The program then passes to the line after 40, which is the end.

These are the three important control statements, though not the only ones. Others include the ON-GOTO statement the if-then-else statement, and the while and until statements. There are also more involved versions of the for-next statements.

In order to solve a large number of mathematics problems, it is necessary to make use of functions. Functions are found almost everywhere in mathematics, and the branch of mathematics known as analysis concentrates most heavily on studying the properties of functions. Many times a programmer will have to construct his own function to do whatever job is needed. But BASIC has a few functions built in that the user can make use of by simply calling the appropriate name (these names are all reserved words, and hence cannot be used for a variable name).

Some of the common functions include the algebraic functions. The function SQR computes the positive square root of a given nonnegative number, and hence it has the format

$$SQR\ (x)$$

where $x \geqslant 0$. The function PI returns the value for pi, $\pi = 3.14159$, and so this function has no argument. The function EXP returns the value of the constant $e = 2.71828$ raised to a power, x; hence the format is

$$EXP\ (x).$$

The logarithm functions LOG and LOG10 represent the natural and common logarithm respectively, and consequently the single argument x must be a positive number. The function ABS returns the absolute value $(y = |x|)$ of any expression x.

Another interesting function is the sign function, SGN, which indicates whether the single argument represents a positive or negative number, or zero. More specifically,

$$SGN\ (x) = \begin{cases} +1 \text{ if x is positive} \\ -1 \text{ if x is negative} \\ 0 \ \text{ if x is zero} \end{cases}$$

so that SGN returns the value $+1$, -1, or 0. The greatest integer function INT, which is most useful in mathematics $(Y=[x])$, returns that integer which is closest to the variable x without ever exceeding it. For instance, INT $(5.1) = 5$, INT $(6.98) = 6$, INT $(14) = 14$, and INT $(-3.4) = -4$. One should note that it always follows that

$$INT\ (x) = x$$

with the equality holding if and only if x is an integer. Closely related to the INT function is the FIX function which returns the truncated value of the argument you supply as a real number. If x is positive then FIX (x) is equal to INT (x). If x is negative, FIX (x) = INT (x) + 1. To illustrate, FIX $(4.1) = 4$, FIX $(8) = 8$, and FIX $(-2.3) = -2$. It follows that

$$FIX\ (x) = SGN\ (x) * INT\ (ABS\ (x))$$

for any real number x.

BASIC supplies the user with four trigonometric functions, namely SIN, COS, TAN, and ATN (arc tangent). It is to be understood that the arguments used with the sine, cosine, and tangent functions are angles expressed in radian measure. Thus the SIN and COS function return values in the range $[-1,1]$, while TAN can return any real number. On the other hand, ATN accepts any real number for the value of its argument, but returns values between $-PI/2$ and $PI/2$.

It is sometimes necessary to have the computer generate a series of random numbers. This is actually impossible to do in the strict sense, but BASIC has been programmed to generate pseudorandom numbers (almost random). The RND function supplies such a series of numbers. This function requires no argument, and each time the RND statement is read, the computer will output a decimal number in the open range 0 to 1. Integral output can be furnished by multiplying RND by the appropriate integer. If we wish to simulate rolling a pair of dice, we could write the following:

```
10   DICE1 = INT (6*RND + 1)
20   DICE2 = INT (6*RND + 1)
30   PRINT DICE1, DICE2
40   END
READY
RUNNH
  3     5
```

Multiplying RND by 6 and then adding 1 produces a decimal somewhere between 1 and 7, then INT returns the integer portion of the number, which in this example gave the values 3 and 5. Each time the computer encounters RND, it returns a different decimal, at least until all seven-place decimals are used up. Unfortunately, each time you run the program above you'll get the same output. BASIC always begins RND with the same value. To alter this starting value, you need only type in a randomize statement (before RND). BASIC will then start the RND function at a new, unpredictable location in the series. This location is sometimes determined by the current time of day according to the computer's clock. Thus, one need only add the single statement

5 RANDOMIZE

to the program above to produce the desired change in starting values for RND.

These functions comprise the typical arithmetical functions possessed by most versions of BASIC. Other functions pertaining to strings, or times or dates can be found in your manual.

BASIC does have one advantage over FORTRAN, and that concerns the ease with which you can handle the arrays that are of two dimensions at the most. The MAT statements in BASIC allow far greater flexibility in handling matrices, whether initializing a matrix, summing two square matrices, computing the inverse, transposition or determinant of a square matrix, or printing the values in an array, than you encounter in FORTRAN. Some of these statements include MAT INPUT, MAT PRINT, MAT READ, INV, TRN and DET. Furthermore, if arrays A and B are given by

$$A = \begin{bmatrix} 1 & 2 \\ 3 & 4 \end{bmatrix}, \quad B = \begin{bmatrix} 5 & 6 \\ 7 & 8 \end{bmatrix}$$

then the statements

```
10   MAT C = A + B
15   MAT D = A * C
```

would quickly set matrices C and D equal to

$$C = \begin{bmatrix} 6 & 8 \\ 10 & 12 \end{bmatrix} \quad D = \begin{bmatrix} 19 & 22 \\ 43 & 50 \end{bmatrix}.$$

Isn't this easy!

No doubt the majority of what has been discussed in the last twenty some pages is rather familiar to you, and well it should if you have some background in computer science. This has just been setting the stage for what is to come: namely, applying BASIC to mathematics; gathering data via an efficient algorithm, testing hypotheses, locating zeros to equations, or approximating complex mathematical expressions. Problems from such areas as number theory, algebra, analysis, and probability and statistics will be presented for you to work on. Many of these problems can be solved in different ways, and it is up to you to be creative in designing methods of attack. Hopefully the variety and nature of the work will stimulate you toward further work in the field. At the very least, you should gain an appreciation for the beauty and purpose of both mathematics and computer science.

Chapter 3
Number Theory

It would be difficult to believe that there is any branch of science that is richer in content, that offers greater variety of problems and types of proofs, that offers more simple problems to stimulate one's interest, and offers more to challenge one's ability than that of number theory. Where else is it possible for an amateur to contribute greatly to the literature on the subject, or for a problem to be so simply stated that everyone can understand it, but no one can solve it? Such is the case with Fermat's problem on solving

$$x^n + y^n = z^n$$

with integral values. And such was the case with the famous four-color problem that was unsolved for some 125 years. The problem was to prove that four colors were sufficient to color any map of states or countries, so that any two neighboring states were colored differently. It wasn't until a few years ago that the problem was finally proved.

Many of the basic theorems of number theory stem from problems investigated by the Greeks, such as perfect numbers, Pythagorean numbers, and primes. Some of the people who contributed early on to the present day body of knowledge include Euclid, Pythagoras, Euler, Fermat, Catalan, Legendre, Descartes, Mersenne, Wilson, Diophantus, Goldbach, and of course, the immortal Karl Frederick Gauss. It was because of these giants that many results were learned; results concerning primality, divisors, residue classes, integral solutions to equations, continued fraction expansions, primitive roots, and representations of integers in various ways. In more recent times, people like Hardy, Littlewood, Baker, Graham, Sierpinski, Selberg, and Erdos have carried on research into the properties and classifications of numbers.

Number theory is an area that lends itself quite well to computer usage. Many results are a function of analyzing ample data, observing patterns, and then verifying theories. The role of the computer is to gather the data, to do all the busywork that used to consume so much time. For instance, Mersenne (Father Marin Mersenne) numbers, M_n, are those numbers of the form

$$M_n = 2^n - 1$$

where n is prime. During the seventeenth century, mathematicians (particularly the French mathematicians, Fermat, Descartes, Desargues, and Pascal) were interested in knowing which M_n were themselves prime. Nine such primes were determined, those corresponding to n = 2,3,5,7,13,17,19,31 and 61. This last prime, M_{61}, is in fact the nineteen digit number 2,305,843,009,213,693,951. It is quite a chore to verify that this number is prime. Euler was unable to do it, but he was able to verify $M_{31} = 2^{31} - 1$ as prime. It wasn't until 1876 that E. A. Lucas was able to show M_{127} was prime [69]. By 1947 mathematicians were able to establish (either by extensive hand calculations, or with desk calculators) the primality of M_{89} and M_{107}. For almost 75 years M_{127} remained the largest Mersenne prime, but, in the early 1950's, Robinson used the SWAC in California and found five new primes, M_{521}, M_{607}, M_{1279}, M_{2203} and M_{2281}. And in 1961 Hurwitz used an IBM 7090 to show M_{4253} and M_{4423} are primes (M_{4423} contains some 2600 digits!!). Computer speed has increased so much since then that in 1978 two high school students, Curt Noll and Laura Nickel, discovered the 25th Mersenne prime. The discovery was national news—Walter Cronkite read the story over CBS-TV. Four months later Noll and David Slowinski found that M_{23209} is prime: the 26th Mersenne prime. And two months after that Slowinski verified that M_{44497} is prime by using a Cray computer [73]. There is no doubt that these results would not have been determined without electronic assistance.

Today, with vastly improved methods and more sophisticated machinery computers are able to do calculations at a much faster rate. Three scientists (Adleman, Rumely, Pomerance) have devised a method [44] that will determine the primality of a 100-digit number within a few hours of computer time. The algorithm can decide if a number n is prime in fewer than $(\log n)^{\log(\log(\log n))^2}$ steps, which is approximately polynomial time. New ways for testing for primes are important to mathematicians because of the side effects brought to light on the harder and more practical problem of factoring. If a factoring algorithm using polynomial time could be discovered, it would have important implications to the government, particularly in the area of cryptography. It was just a few years ago at MIT that Aldeman, Ronald Rivest, and Adi Shamir had constructed [30] an "unbreakable" code, a code that the government was most interested in. The principle behind the code centered around the use of a 125-digit number and its two prime factors. Knowledge of the factors would be needed to break the code. Rivest estimated that if the fastest computer were employed, using the most efficient algorithm known, it would take roughly 40 quadrillion years to determine the two prime factors. This is certainly long enough to say that the code is unbreakable! Unfortunately, for some, the most recent methods may break the code. Mathematics has thus become a question of national security.

The topics that appear in this chapter have been especially selected because it will not be difficult to program the gathering of data and the search for solutions to them, and because the topics are interesting and challenging enough to prove worthwhile.

SIMPLIFYING FRACTIONS

We all know that 9/12 is equivalent to 3/4, and that 21/56 is equivalent to 3/8. Nothing is profound about either of these statements. But it is somewhat interesting to note that 16/64 is equal to 1/4, which, a careless student may surmise using the following reasoning

$$\frac{1\cancel{6}}{\cancel{6}4} = \frac{1}{4}$$

where the two sixes are erroneously cancelled out. This bit of blind luck is not reserved for this particular example alone. The fraction 19/95 is equal to 1/5, and again we have the faulty cancellation,

$$\frac{1\cancel{9}}{\cancel{9}5} = \frac{1}{5}$$

Also, 26/65 equals 2/5, which follows if the two sixes are removed. There is one other (four altogether) example of a proper fraction, with two digits in both the numerator and denominator, for which this false cancellation works. Can you find it?

You could proceed by examining fractions with larger numerators and denominators, say three or more digits each. You would find more examples where a digit, or perhaps two digits, could be cancelled out, leaving an equivalent fraction. This would happen with 484/847 = 4/7, where both a 4 and an 8 could be cancelled. This, however, is a direction that I will not pursue in this section. You may elect to explore this fertile ground on your own, but for the remainder of this section, I wish to consider just those proper fractions with two-digit numerators and denominators.

For the sake of simplifying notation, let us agree to write any two digit number tu (the t stands for tens, and the u for units) in the form

$$t,u$$

where the comma is used to separate the digits. The reason for this will soon become apparent.

Suppose we wish to write a program that determines all the proper rational fractions where the faulty cancellation law holds. This means we want those fractions where either

$$\frac{t,u}{u,w} = \frac{t}{w}$$

or

$$\frac{t,u}{w,t} = \frac{u}{w} \, .$$

So either the units digit in the numerator is the digit that gets cancelled, or the tens digit gets cancelled. These two problems are equivalent, because if the tens digit t in the numerator gets canceled,

$$\frac{\cancel{t},u}{w,\cancel{t}} = \frac{u}{w}$$

then you would have also had

$$\frac{w,\cancel{\ell}}{\cancel{\ell},u} = \frac{w}{u}$$

where the units digit in the numerator is cancelled. The problem can be refined as follows:

Find the digits t,u,w with $t < u$ so that

$$\frac{t,u}{u,w} = \frac{t}{w} .$$

This way we agree to cancel only the units digit of the numerator with the tens digit of the denominator. Sample output would then show $19/95 = 1/5$ instead of $95/19 = 5/1$.

From a programming standpoint, how would you find integers t,u,w so that $t < u$, and t,u/u,w equals t/w? One should keep in mind that the numerator t,u is simply a symbolic way of expressing the integer $10*t + u$. Likewise, u,w represents $10*u + w$. The following program might provide an adequate search.

Program

```
100   FOR T = 1 TO 8
110     FOR U = (T + 1) TO 9
120       FOR W = 2 TO 9
130         FRACTION1 = (10 T + U)/(10 U +
            W)
140         FRACTION2 = T/W
150         IF FRACTION 1 ≠ FRACTION 2 GO
            TO 170
160         PRINT "WE HAVE A SOLUTION. T
            = ";T;"U = ";U;"W = ";W
170         NEXT W
180     NEXT U
190   NEXT T
200   STOP
```

This would work well if we didn't have to worry about roundoff error when computing the two fractions, t,u/u,w and t/w. It might be better if you cross multiply, thereby having to only equate integers. Lines 130, 140, and 150 from the program could be replaced by,

```
130   INTEGER1 = W*(10*T + U)
140   INTEGER2 = T*(10*U + W)
150   IF INTEGER1 ≠ INTEGER2 GO TO 170
```

In an effort to add some content to this problem, we change all
our arithmetic from a base 10 system to base b, where b ≥ 2.
Nothing of interest is derived from b = 2 or b = 3, because there just
aren't enough digits. But base 4 does produce an interesting result
since

$$\frac{1,3}{3,2} = \frac{1}{2}.$$

Keep in mind that 1,3 represents the number $1 \cdot b^1 + 3 \cdot b^0 = 1 \cdot 4 + 3$
= 7, and 3,2 represents $3 \cdot b + 2 = 14$, hence 7/14 = 1/2. This
example can serve to spur you on and find other examples in base b,
and perhaps you can discover some developing patterns for various
bs.

First, you need to reformulate the problem. For a given b ≥ 4,
seek values t,u,w where t,u, and w are digits in base b (this means
t,u,w ≤ b − 1) with t < u and

$$\frac{t,u}{u,w} = \frac{t}{w}.$$

This last equality can be interpreted to mean

$$\frac{tb + u}{ub + w} = \frac{t}{w}.$$

or, equivalently, w(tb + u) = t(ub + w). Now that the problem has
been restated, you can consider how to solve it. It shouldn't be that
much different from the earlier case involving base 10. Wherever
the digit 10 appeared before, the value of b will take its place.
Likewise, b − 1 will replace the 9, and b − 2 will replace the 8. You
should consider that, if b > 10, some of the digits in the base b
system will be digits with more than one base 10 digit. Thus, in base
14, the digits are 0, 1, 2, •••, 9, 10, 11, 12, 13. The number in base 14
represented by 12,11 is the number equivalent to $12 \cdot 14 + 11$, which
is 179 in base 10. The comma between the 2 and the 1 serves to

distinguish between the two digi
number

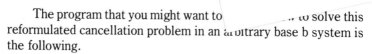

$$\frac{3,15}{15,13}$$

is equivalent to 3/13 (we cancel the

$$3[3(26) + 15] = 13[15$$

The program that you might want to　　　　　　.. to solve this reformulated cancellation problem in an arbitrary base b system is the following.

Program

```
100   INPUT B
110   FOR T = 1 TO (B − 2)
120     FOR U = (T + 1) TO (B − 1)
130       FOR W = 2 TO (B − 1)
140         INTEGER1 = W*(B*T + U)
150         INTEGER2 = T*(B*U + W)
160         IF INTEGER1 ≠ INTEGER2 GO TO
              180
170           PRINT "WE HAVE A SOLUTION. T =
              ";T;"U = ";U;"W = ";W
180       NEXT W
190     NEXT U
200   NEXT T
210   STOP
```

This program is short and simple, and could certainly locate all the solutions in a brief amount of time, especially if b is small. What if b were to equal 1000? The number of combinations of t, u, and w is just shy of 500,000,000. This corresponds to the total number of fractions

$$\frac{t,u}{u,w}$$

to be considered. Considerable computer time would be consumed in checking all of these fractions to see if the digit u could be cancelled. But, a little investigation into the problem reveals some time saving shortcuts.

program occur when the equation $w(tb + u) =$ [...] [...]ed. Rearranging this gives $bt(u - w) = w(u - t)$. [...] must divide both sides of the equation, and if b [...] be prime, it follows that either b divides w (written $b|w$) [...] $(u - t)$. Both of these are impossible since both w and $(u - t)$ [...] smaller than b. Consequently there is no need to investigate those cases when $b = 2, 3, 5, 7, 11, 13, \cdots$.

Additional insight into this problem is provided in [8] where it is shown that if b is not prime, b will be of the form $b = k_1 k_2$ with $1 < k_i < b$, and there is a solution, which is of the form

$$t = (b/k_i) - 1$$
$$u = b - 1$$
$$w = b - k_i.$$

This may not be the only solution, but there are at least as many solutions as there are proper factors of b. We can illustrate this by setting $b = 28$. Then the proper factors are 2, 4, 7, and 14. There must be a minimum of four solutions to the problem, these being

a. $t = 13$	b. $t = 6$	c. $t = 3$	d. $t = 1$
$u = 27$	$u = 27$	$u = 27$	$u = 27$
$w = 26$	$w = 24$	$w = 21$	$w = 14$

There are actually six more solutions in this case, but this would have been the exact number of solutions had $27 = b - 1$ been prime.

The reference [8] adds further shortcuts to the programming. First, the units digit u is never allowed to be too small. Its minimum size is determined by b through the relationship

$$u \geq 1 + \sqrt{b}.$$

The digit w is also restricted from being too small. (These restrictions on both w and u are observed in the four solutions listed above.) Its minimal size is controlled by u, with

$$w \geq \frac{u + 1}{2}.$$

The tens digit is not allowed to be too large, where its maximum value is bounded by

$$t \le \frac{w}{2}.$$

Because of these restrictions, perhaps the program should be altered so that the loop on t is the innermost loop, and the loop on u is the outermost loop.

```
105   LOWER1 = INT(1 + SQR(B))
110   FOR U = LOWER1 TO (B − 1)
115       LOWER2 = INT((U + 1)/2)
120       FOR W = LOWER2 TO (B − 1)
125           UPPER = INT(W/2)
130               FOR T = 1 TO UPPER
```

One final result perhaps best sums up how to quickly locate the appropriate values for t, u, and w. This result will drastically reduce the computer time needed to search through all possible quotients of two-digit numbers in base b.

RESULT: All solutions t, u, w are of the form

$$u = mp$$
$$w = u - k$$
$$t = uw/(w + kb)$$

where p is a prime factor of b − 1, m is a number no greater than (b − 1)/p, and k is a number that is strictly less than u.

Thus, if you wanted to try and find the other six solutions when b = 28, you would first note that the prime factor of b − 1 is p = 3. So u must be a multiple of 3, with u = 3m and $1 \le m \le 9$. The possibilities must be u = 3, 6, 9, 12, 15, 18, 21, 24, 27. If you try u = 9, w must be of the form w = 9 − k, with k < 9. Then for t you have

$$t = \frac{uw}{w + kb}$$

$$= \frac{9(9 - k)}{(9 - k) + 28k}$$

$$= \frac{9 - k}{3k + 1}$$

45

and the only way t can be a positive integer is when k = 1 (t = 2) or k = 2 (t = 1). Thus, u = 9, w = 8, t = 2 and u = 9, w = 7, t = 1 provide two more solutions when b = 28.

Exercises

1. Write a program that finds all solutions to

$$\frac{t,u}{u,w} = \frac{t}{w}$$

where t, u, w are digits in a base b system, with t < u. Assume b ≤ 100. Test your program with the following values of b.

b = 28 (10 solutions)
b = 37 (0 solutions)
b = 100 (37 solutions).

2. For which values of b are there precisely one solution to

$$\frac{t,u}{u,w} = \frac{t}{w} \quad , t < u?$$

NEGATIVE BASES

A standard and well known property of our number system is the unique representation of each positive integer in an integral base system other than 10. More specifically, if b is some integer, at least as great as two, then any positive integer N can be expressed in the form

$$N = \sum_{i=0}^{m} a_i b^i$$

where the a_i (selected from the set of digits in the base b system) can assume any value 0, 1, 2, ⋯, (b − 1). This means that if we wish to express N = 37 in the base 4 system, we would have to find those digits for a_0, a_1, and a_2 with 0 ≤ a_i ≤ 3 and

$$37 = a_2(4^2) + a_1(4^1) + a_0$$
$$= 16a_2 + 4a_1 + a_0.$$

Trial and error produces the desired findings: we set $a_2 = 2$, $a_1 = 1$ and $a_0 = 1$. Thus, the number 37, in base 10 (our decimal system), can be written as 211 in base 4. We write this as $37_{10} = 211_4$. Other examples would include,

$$73_{10} = 1{\cdot}4^3 + 0{\cdot}4^2 + 2{\cdot}4^1 + 1 = 1021_4$$
$$100_{10} = 1{\cdot}4^3 + 2{\cdot}4^2 + 1{\cdot}4^1 + 0 = 1210_4$$
$$409_{10} = 3{\cdot}4^4 + 2{\cdot}4^3 + 1{\cdot}4^2 + 2{\cdot}4^1 + 1 = 32121_4.$$

Some examples from base 5 (keep in mind that the digits are 0, 1, 2, 3, 4) include

Int digits less than 5

$$73_{10} = 2{\cdot}5^2 + 4{\cdot}5^1 + 3 = 243_5$$
$$100_{10} = 4{\cdot}5^2 + 0{\cdot}5^1 + 0 = 400_5$$
$$409_{10} = 3{\cdot}5^3 + 1{\cdot}5^2 + 1{\cdot}5 + 4 = 3114_5.$$

It is really not difficult to change a number N from its base 10 representation to its equivalent representation in base b. We first find the highest power of b that divides into N, giving a quotient that is between 1 and $b - 1$ in value, inclusively. Thus we need to find m such that

$$1 \leqslant \frac{N}{b^m} \leqslant (b - 1).$$

In the particular case with $N = 73$ and $b = 5$ we have $73/b^0 = 73$, $73/b^1 = 14.6$, and $73/b^2 = 2.92$; this last figure of 2.92 is between 1 and $b - 1 = 4$. Once you have found this m, set a_m equal to the integral portion of the quotient; thus

$$a_m = INT\ (N/b^m).$$

With $N = 73$ and $b = 5$, you have $m = 2$ and $a_2 = 2$. This means that the representation of N in base b takes the initial form

$$N_{10} = a_m b^m + a_{m-1} b^{m-1} + \cdots + a_0$$
$$= a_m a_{m-1} - - - a_0{}_b$$

$$\underbrace{\qquad\qquad}_{m + 1\ \text{digits}}$$

where you know the value of a_m. To compute the next digit, a_{m-1}, you could reset N to be $N - a_m b^m$. Then a_{m-1} will be the integral portion (and it could be zero) of the quotient of $(N - a_m b^m)/b^{m-1}$.

$$a_{m-1} = INT \left[\frac{N - a_m b^m}{b^{m-1}} \right].$$

If we do reset $N = N - a_m b^m$ then it would follow that

$$a_{m-1} = INT(N/b^{m-1}).$$

This process continues

$$N = N - a_{m-1} b^{m-1}$$

$$a_{m-2} = INT(N/b^{m-2})$$
$$\bullet$$
$$\bullet$$
$$\bullet$$

until a value for a_1 is determined. Then the final digit, a_0, is the difference $N - a_1 b^1$. In each case, the value of N is the reset value of the previous N minus the appropriate multiple of b.

It is an even easier problem to convert a number from its representation in base b to its representation in base 10. For if the number 234_6 is input data (this is 234 in base 6), the equivalent base 10 representation is $2 \cdot 6^2 + 3 \cdot 6^1 + 4$, or 94. In general, if you are given the base b number $a_m a_{m-1} \cdots a_0$, then its base 10 representation is

$$a_m b^m + a_{m-1} b^{m-1} + \cdots + a_0.$$

College courses entitled "Mathematics for Elementary Education" typically contain material on variable base arithmetic, so that these teachers-to-be will better comprehend the mechanics of our base 10 system. Students of computer science are more familiar with the two special cases when $b = 2$ (binary system) and $b = 16$ (hexadecimal system). Interestingly enough, the binary system yields some interesting applications to Mathematical Game Theory, as exemplified by the ancient game of Nim, and more recently, by the games of Rims, and the Silver Dollar Game Without the Dollar [29].

Mathematicians love to generalize a theory or a result as much as possible. Variable base arithmetic has proved a fruitful area for doing just this. One such generalization [78] was to allow for negative digits in the expansion of N in base b. For instance, when setting b = 5, instead of requiring the digits a_i to assume only the values 0, 1, 2, 3, 4, it is possible to make worthwhile results from requiring a_i to assume only the values $-3, -2, -1, 0, 1$. When this is the case, you could express N = 73 as

$$73_{10} = 1(5^3) - 2(5^2) + 0(5^1) - 2 = 1(-2)(0)(-2)_5$$

The number N = 11 could be expressed as $11_{10} = 1(-3)(1)_5$ because

$$11 = 1(5^2) - 3(5) + 1.$$

Any set of five consecutive integers containing both 0 and 1 will serve as an adequate set of digits for a base 5 system. The set $\{-2,-1,0,1,2\}$ would therefore also suffice. In general, for base b, any set of b consecutive integers containing both 0 and 1 will act as a proper set of digits. In base 3 we could elect to choose our digits from the set $\{-1,0,1\}$. This situation will be discussed, and used, in the next section on Bachet's weights.

A futher generalization of variable base number systems is to allow the base b to be a negative integer, b < 0. This change can produce some interesting results, as you shall see shortly. Again, you first need to consider what your digits are. Even though you could use any |b| consecutive integers that include both 0 and 1, for the sake of simplicity, confine yourself to the standard set of S = $\{0,1,2,\cdots,|b| - 1\}$. It follows [31] that any positive integer N can be uniquely represented as a finite sum of multiples of powers of b,

$$N = \sum_{i=0}^{m} a_i b^i$$

where $a_i \in S$.

If you set b = -10, the set S of digits is $\{0,1,2,\cdots,9\}$. If you choose N = 73, you would have to find that values of a_i so that

49

$$73 = a_4(-10)^4 + a_3(-10)^3 + a_2(-10)^2 + a_1(-10)^1 + a_0$$
$$= 10000a_4 - 1000a_3 + 100a_2 - 10a_1 + a_0.$$

Since

$$73 = 70 + 3$$
$$= (100 - 30) + 3$$
$$= 100(1) - 10(3) + 3$$

it follows that $a_4 = 0$, $a_3 = 0$, $a_2 = 1$, $a_1 = 3$ and $a_0 = 3$. You can omit a_4 and a_3 from your representation since they are equal to zero. Consequently

$$73_{10} = 133_{-10}.$$

Similarly, with N = 274, you find

$$274 = 300 - 30 + 4$$
$$= 100(3) - 10(3) + 4$$
$$= 334_{-10}$$

and for N = 918 you have

$$918 = 1000 - 90 + 8$$
$$= 10000 - 9000 - 90 + 8$$
$$= 10000(1) - 1000(9) + 100(0) - 10(9) + 8$$
$$= 19098_{-10}.$$

You might be able to deduce a pattern evolving for representing N in base -10. If $N = n_1 n_2$ is any two-digit number, then

$$N = n_1 n_{2_{10}} \begin{cases} = n_1(10) + n_2 \\ = 100 - 10(10 - n_1) + n_2 \\ = 100(1) - 10(10 - n_1) + n_2 \\ = 1(10 - n_1)n_{2_{-10}}. \end{cases}$$

But if $N = n_1 n_2 n_3$ is any three-digit number, we have

a) if $n_1 < 9$, $n_1 n_2 n_{3_{10}} = n_1(100) + n_2(10) + n_3$

50

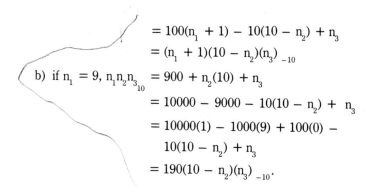

$$= 100(n_1 + 1) - 10(10 - n_2) + n_3$$
$$= (n_1 + 1)(10 - n_2)(n_3)_{-10}$$

b) if $n_1 = 9$, $n_1 n_2 n_{3_{10}} = 900 + n_2(10) + n_3$
$$= 10000 - 9000 - 10(10 - n_2) + n_3$$
$$= 10000(1) - 1000(9) + 100(0) -$$
$$10(10 - n_2) + n_3$$
$$= 190(10 - n_2)(n_3)_{-10}.$$

You should try to express any four-digit number $N = n_1 n_2 n_3 n_4$ in base (-10). How about $N = n_1 n_2 n_3 \cdots n_k$?

Consider the following question: Which number is larger, 23_{-10} or 32_{-10}? Because $23 = 2$ $(10) + 3 = -17$ and $32 = 3$ $(10) + 2 - 28\ 10$, it follows that 23_{-10} is the larger. How about 230 or 320_{-10}? Which is larger? It has to be 320_{-10} because $320_{-10} = 3(-10)^2 + 2(-10) + 0 = 280_{10}$, while $230_{-10} = 170_{10}$. When comparing two numbers in base b, it is important to note whether they have an odd number of digits (the leading exponent of b is even, so the leading power is positive) or an even number (the leading exponent of b is odd, so the leading power is negative).

Consider a different negative base, say, $b = -3$. This time the digits come from the set $S = \{0,1,2\}$, and any positive integer N can be represented as

$$N = \sum_{i=0}^{m} a_i(-3)^i$$

with $a_i \in S$. To illustrate for $N = 37$,

$$37_{10} = 81 - 54 + 9 + 1$$
$$= 1(-3)^4 + 2(-3)^3 + 1(-3)^2 + 0(-3)^1 + 1$$
$$= 12101_{-3}$$

and for $N = 61$,

$$61_{10} = 81 - 27 + 9 - 3 + 1$$
$$= 1(-3)^4 + 1(-3)^3 + 1(-3)^2 + 1(-3)^1 + 1$$
$$= 11111_{-3}.$$

Consider how to systematically determine the base -3 representation of N. You need an algorithmic approach to determine the values of a_i. Should you first compute the value of m, the highest power of b? It is clear that, since N is positive, the value of m must be even. Perhaps you may wish to find this m; or perhaps you note that it might be feasible to express N as the sum of a constant, plus odd powers of $|b|$,

$$N = c_0 + c_1|b|^1 + c_3|b|^3 + c_5|b|^5 + \cdots + c_{2k+1}|b|^{2k+1}$$

where $0 \leq c_i \leq |b|$. Note that with $N = 61$ and $b = -3$ you have

$$61 = 1 + 6 + 54$$
$$= 1 + 2 \cdot 3^1 + 2 \cdot 3^3$$

where $c_0 = 1$, $c_1 = 2$ and $c_3 = 2$. This proves useful because

$2 \cdot 3^3 = (3-1)3^3 = 3^4 - 3^3 = (-3)^4 + (-3)^3$ and $2 \cdot 3^1 = (3-1)3^1 = 3^2 - 3^1 = (-3)^2 + (-3)^1$. Consequently, $61 = (-3)^4 + (-3)^3 + (-3)^2 + (-3)^1 + 1 = 11111_{-3}$. The principle here is that

$$c_{2k+1}|b|^{2k+1} = \begin{cases} (|b| - c_{2k+1})|b|^{2k+1} & \text{if } c_{2k+1} < |b| \\ b^{2k+2} & \text{if } c_{2k+1} = |b| \end{cases}$$

$$= \begin{cases} b^{2k+2} + c_{2k+1}b^{2k+1} \\ b^{2k+2} \end{cases}$$

$$= \begin{cases} a_{2k+2}b^{2k+2} + a_{2k+1}b^{2k+1} & \text{with } a_{2k+2} = 1, \text{ and} \\ & \qquad\qquad a_{2k+1} = c_{2k+1} \\ a_{2k+2}b^{2k+2} & \text{with } a_{2k+2} = 1, \text{ and} \\ & \qquad\qquad a_{2k+1} = 0. \end{cases}$$

In other words, if you know the values of c_i, you can easily compute the coefficients a_0, a_1, $\cdots$, a_m. You should experiment with the base representations and try to find a suitable pattern—after all, this is really part of the learning process.

The final concern in this section involves some basic arithmetic in a negative base system. For instance, if you want to add the two numbers $x_1 = 204_{-10}$ and $x_2 = 108_{-10}$, you must first add the two digits, 4 and 8, in the units column. The total is 12_{10}, which is equivalent to 192_{-10}. Hence you must write down the 2, and carry the 19 (one digit to a column).

$$
\begin{array}{r}
19 \\
204_{-10} \\
+\ 108_{-10} \\
\hline
2
\end{array}
$$

The digits 9 and 0 sum to 9_{10}, which is equivalent to 9_{-10}. This gives you a total of 492_{-10}.

$$
\begin{array}{r}
19 \\
204_{-10} \\
+\ 108_{-10} \\
\hline
492_{-10}
\end{array}
$$

On the other hand, summing 27_{-10} and 36_{-10} gives a carry of 19

$$
\begin{array}{r}
19 \\
19 \\
19 \\
27_{-10} \\
+\ 36_{-10} \\
\hline
\cdots 0043_{-10}
\end{array}
$$

an infinite number of times. An infinite string of zeros appears in the sum, preceding the digit 4. This sum can be interpreted as 43_{-10} where you just omit the zeros.

This particular feature of having the carry digits accumulate in an infinite string is not an unusual occurrence, and it makes no difference which negative base you are in. You are referred to [31] for more details on negative base arithmetic.

Exercises

1. Write a program which gives the representation of N (a given positive integer from base 10) in base b (a given negative integer). Test your program with the following.

Input		Output
N	b	
73	−10	133
274	−10	334
918	−10	19098
37	−3	12101
61	−3	11111
12	−2	11100
11	−2	11111
59	−15	1(12)(14)

2. If $x_1 = 19_{-10}$ and $x_2 = 1_{-10}$ and $x_3 = 23_{-10}$ then find
   a. $x_1 + x_2 + x_3$
   b. $x_1 + x_2$
   c. $x_1 \cdot x_3$

BACHET'S WEIGHTS

There seems to be a certain class of problem that is commonly found in puzzle or amusement books. This class contains a number of classic number theory problems. One of these is the challenge of determining the single counterfeit coin out of a given collection using only a limited number of weighings on a pan balance. More specifically, you have the following two problems.

Problem 1: Among 12 similar coins there is precisely one counterfeit. It is not known whether the counterfeit coin is lighter or heavier than a genuine one. (All genuine coins weigh the same.) Using three weighings on a pan balance, how can the counterfeit coin be identified and in the process, be determined to be lighter or heavier than a genuine coin?

Problem 2: You have eight similar coins, and a beam or pan balance. At most one coin is counterfeit and hence underweight. How can you detect whether there is an underweight coin, and if so, which one, using the balance only twice?

In both these problems and in others of similar type [51], one collection of coins is initially placed on one end of the balance, and an equal number of coins is placed on the other end of the balance. If the two collections balance, all those coins are genuine and it is known that the counterfeit coin must be in among the remaining coins. A second weighing then takes place, using two more appropriately chosen distinct collections of coins. If the original two collections do not balance, the faulty coin is present in one of these two collections, thus providing knowledge as to how to select the next two collections to be weighed.

Now direct your attention to a different kind of weighing problem. More specifically, there is an object of unknown weight, say W. The weight is assumed to be integral and no greater than 100; so W $\in$ {1,2,•••,100}. You wish to determine the value of W. You have a set of standard weights that you place on one end of a balance, and you place the object on the other. Now certainly, if you had enough different standard weights, the weight of any unknown object could be determined. For instance, weights of 1, 2, 3, 5, 6, 7, 9, 10, 11, 13, 14, 15, 17, •••, 47, 48, 49 can be used to weigh any object up to 100. These weights are all the integral weights from 1 to 50 except for the multiples of 4. Likewise, all the integral weights from 1 to 50 except for the multiples of 5 and 6 can be used to weigh any object up to 100. Table 3-1 lists some more sets of weights that can function in this capacity.

According to the last entry in the table, ten standard weights are sufficient to weigh any object up to 100. If W = 34, for instance, the standard weights 1, 3, 4, 5, 6, 7, 8 would be used. See Fig. 3-1. If W = 60, the weights 2, 6, 7, 8, 37 suffice.

The important question is whether or not you could get by with fewer than ten standard weights. Can you find nine or eight standard

Table 3-1. Various Standard Weights.

Standard Weights	Total Number of Standard Weights
1,2,3,•••,10,21,22,•••,30, 41,42,43,•••,50	30
1,2,3,•••,15,41,42,•••,50	25
1,1,2,2,3,3,•••10,10	20
1,2,3,•••,13,14,15	15
1,2,3,•••,7,8,37,74	10

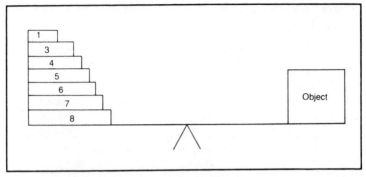

Fig. 3-1. Obtaining W = 34 with standard weights.

weights that when placed on one end of a beam balance in the right combination will balance with any object of integral weight not exceeding 100 placed on the other end? As it turns out, a minimum of seven weights will perform this task. The weights 1, 2, 4, 8, 16, 32, 64 will weigh any object up to 100. This set of weights is not unique, because 1, 2, 4, 8, 16, 32, k ($37 \leq k \leq 64$) will also suffice. The set 1, 2, 4, 8, 16, 32, 64 actually weighs any object up to 127, and in this instance, this is the only set of 7 weights to do this. In general, the standard weights 1, 2, 4, 8, •••, 2^n weigh any amount W with $W \leq 2^{n+1} - 1$. Furthermore, if $W = 2^{n+1} - 1$, these $(n + 1)$ weights are the only (and hence the fewest) $n + 1$ weights to weigh any object of weight W. This happens because every positive integer has a unique representation in the binary (base 2) system. Thus, if an object has weight W, where W has a binary representation of

$$W = \sum_{i=0}^{n} a_i 2^i$$

then, at most, $n + 1$ standard weights $a_i 2^i$ will sum to W.

Now consider a slight variation of this problem. Instead of placing the standard weights on the end of the balance opposite the object, place the weights on both ends: some weights on one end and some on the other. So again, if $W = 100$, what is the minimum number of weights (and what are they) necessary to balance the object of weight W and determine W? In this situation you are able to reduce the number of standard weights to five. The necessary weights are 1, 3, 9, 27, and 81. To show how this works, use $W = 45$. Place the two standard weights 9 and 27 on the same end of the balance as the object and the 81 weight on the opposite end as shown in Fig. 3-2.

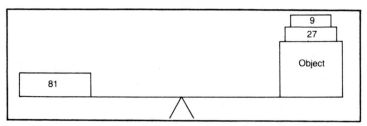

Fig. 3-2. Obtaining W = 45 with weights on both ends of the balance.

Because the standard weights are powers of three, it is reasonable to assume that the placement of the standard weights is dictated by the ternary (base 3) representation of W. And this is the case, though it is not obvious what the specific relationship is. For, representing 45 in base 3 gives

$$45 = 27 + 18$$
$$= 1(3^3) + 2(3^2) + 0(3^1) + 0(3^0)$$
$$= 1200_3$$

Nothing about this expression suggests that the 9 and 27 weights should be placed on the same end of the balance as the object, or that the 81 weight be placed at the opposite end. Here you must make use of one of the generalized base representations as was discussed in the previous section. You do want to express W in base 3, but not with the customary digits 0, 1, 2. Instead use the three digits -1, 0, 1. In this way you have

$$45 = 81 - 27 - 9 + 0 + 0$$
$$= 1(-1)(-1)00_3$$
$$= a_4 a_3 a_2 a_1 a_{0_3}.$$

From this you can see that the two negative digits, a_3 and a_2, indicate which standard weights are on the same end of the balance as the object. Thus, the positive digits indicate which weights are on the opposite end of the balance.

Let's try a few more examples.

Example 1: Set W = 22.

Since $22 = 27 - 9 + 3 + 1 = 1(-1)11_3$, put the 9 weight with the object, and the 1, 3, 27 weights on the opposite end, as shown in Fig. 3-3.

Example 2: Set W = 70.

Since $70 = 81 - 9 - 3 + 1 = 10(-1)(-1)1_3$, put the 3 and 9 weight with the object, and the 1 and 81 weights on the opposite end.

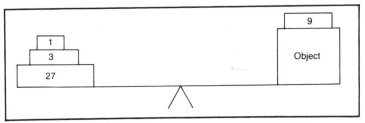

Fig. 3-3. Obtaining W = 22 with weights on both ends of the balance.

You are going to write a program that simulates this process, which is known in the literature as Bachet's problem of the weights [36]. The program must take the input value for W, obtain its base 3 representation using the digits $-1, 0, 1$, and then assign the proper powers of three to the appropriate balance end. In other words, once W is fed into the program, you must first compute the base 3 representation

$$W = \sum_{i=0}^{n} a_i 3^i$$

with $a_i \in \{-1, 0, 1\}$. Then, if $a_i = 1$, the standard weight 3^i is placed opposite the object, while if $a_i = -1$, the 3^i weight is placed on the same end of the balance as the object. A weight W equal to $1(-1)(-1)011_3$ would therefore have the two standard weights 27 and 81 alongside it, and the three weights 1, 3, 243 opposite it.

Since the program is basically divided into two main portions, namely obtaining the base 3 representation and positioning the individual standard weights, take a look at the first portion and see what is involved. The second portion shall be left to the reader.

If you decide to store the digits a_i in an array, say A, W will be represented by

$$W = A(n)A(n-1)\cdots A(1)A(0)_3$$

where each $A(i)$ is either 0, 1, or -1. You need to first compute n, the largest subscript of the array. Note that $n = 3$ for $1(-1)11_3$, while $n = 5$ for $1(-1)0(-1)11_3$. To this end, define two more arrays, one named Q (for quotient), and one named R (for remainder). The quotient terms, written mathematically as q_i, and the remainder terms, r_i, are integers that satisfy the conditions

1. $q_i \in \{-1,0,1\}$
2. $0 \leqslant |r_i| \leqslant 1 + 3 + 3^2 + \cdots + 3^{i-1}$

(Note: this sum is equal to $(3^i - 1)/2$.)

To find n, seek the largest value of i for which $W = 3^i q_i + r_i$. The value a_n will be set equal to q_i. Consider $W = 58$; the largest i for which $58 = 3^i q_i + r_i$ is $i = 4$, because we let $q_4 = 1$ and $r_4 = -23$ $(|r_4| = 23 \leqslant (3^4-1)/2)$,

$$58 = 3^4(1) - 23$$

so $a_4 = q_4 = 1$. Thus, we know our base 3 representation for W is of the form

$$58 = a_4 a_3 a_2 a_1 a_0$$
$$= 1a_3 a_2 a_1 a_0$$
$$= A(4)A(3)A(2)A(1)A(0)_3.$$

Next, to find a_3, consider the equation

$$r_4 = 3^3 q_3 + r_3$$

or equivalently

$$-23 = 3^3 q_3 + r_3.$$

In this case you can choose $q_3 = -1$ and $r_3 = 4 (|r_3| = 4 \leqslant (3^3-1)/2 = 13)$. You then set $a_3 = q_3 = -1$, and this gives

$$58 = a_4 a_3 a_2 a_1 a_0$$
$$= 1(-1)a_2 a_1 a_0.$$

Continue the pattern with

$$r_3 = 3^2 q_2 + r_2$$
$$r_2 = 3^1 q_1 + r_1$$
$$r_1 = 3^0 q_0 + r_0$$

and if the calculations are correct the last remainder, r_0 will equal zero. The remaining coefficients a_2, a_1, a_0 will be equal to q_2, q_1, q_0 respectively. Suppose that r_{i+1} has been determined, and you are then looking at the equation

$$r_{i+1} = 3^i q_i + r_i.$$

What is the best way to program to solve for q and r_i? First, if r_{i+1} is positive, set $q_i = 1$, and if r_{i+1} is negative, set $q_i = -1$. Then check whether $r_i = r_{i+1} - 3^i q_i$ has an absolute value no greater than $(3^i - 1)/2$. If this holds, you are done. You can set $a_i = q_i$ and then move on to the next equation $r_i = 3^{i-1} q_{i-1} + r_{i-1}$. If the absolute value of r_i is greater than $(3^i - 1)/2$, set $q_i = 0$ and $r_i = r_{i+1}$, and move on to the next equation $r_i = 3^{i-1} q_{i-1} + r_{i-1}$. Continue this process until $i = 0$. This form of the Euclidean algorithm will always work, and it provides the systematic algorithm that programmers strive to find.

As a last example we set $W = 299$; the string of equations that follow is

$$299 = 3^5(1) + (56)$$
$$56 = 3^4(1) + (-25)$$
$$-25 = 3^3(-1) + (2)$$
$$2 = 3^2(0) + (2)$$
$$2 = 3^1(1) + (-1)$$
$$-1 = 3^0(-1) + (0)$$

It follows that n = 5, and $a_5 = 1$, $a_4 = 1$, $a_3 = -1$, $a_2 = 0$, $a_1 = 1$ and $a_0 = -1$; or

$$299 = 11(-1)01(-1)_3.$$

If an object that weighs 299 is placed on one end of the balance, the two standard weights of 1 and 27 would be placed with it, and the three weights of 3, 81, and 243 placed on the opposite end would balance it.

Exercises

1. Write a program that converts the input value W into its base 3 representation, with digits 0, ± 1, and then determine which standard weights 1, 3, 9, 27, ••• should be placed on the appropriate end of the balance. Assume W is no greater than 10,000. Test your program using W = 8247.
2. Solve Problem (1) as stated in the beginning of the section.

LATTICE POINTS

Lattice points are those points (x,y) in the plane (that is if you wish to work in 2-dimensional geometry; otherwise you consider 3-space or higher) where both coordinates, x and y, are integral. Thus, the points (−1,1), (0,3), (1417,−81) are lattice points, while (3,.7), ($\sqrt{2}$,1) and (1/2,18) are not. The locus of all lattice points as shown in Fig. 3-4 forms a lattice type grid in the plane.

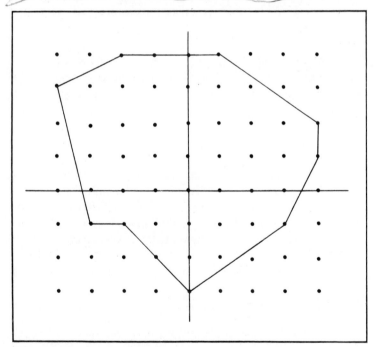

Fig. 3-4. A simple lattice polygon.

There have been some very important mathematical results established regarding lattice points. Integral solutions to linear equations, otherwise known as Diophantine equations, correspond to those lattice points in the plane that the graph of the function passes through. These kinds of numerical solutions are sometimes the only meaningful ones because of the nature of the variables. Geometry also yields some interesting results. One such statement is known as Pick's Theorem [49]. George Pick discovered a most elegant formula for the area of a simple lattice polygon. A simple lattice polygon is one in which the vertices are described by coordinates that are integers, and the line segments that form the sides do not intersect anywhere other than at the vertices. Figure 3-4 illustrates a simple lattice polygon.

According to Pick, the area is completely determined by the lattice points that are interior to the polygon and those that fall on the boundary (the line segments) of the polygon. If I denotes the number of interior lattice points, and B denotes the number of boundary lattice points, Pick's formula reads,

$$A = I + \frac{B}{2} - 1.$$

The polygon sketched in Fig. 3-4 contains $I = 33$ interior lattice points, and $B = 12$ boundary lattice points. These 12 points include (0,4), (1,4), (4,2), (4,1), (3,−1), (0,−3), (−1,−2), (−2,−1), (−3, −1), (−4,3), (−2,4), and (−1,4). This gives a polygonal area of $A = 33 + 12/2 - 1 = 38$.

The main point of discussion for this section centers around a problem that was mentioned by Paul Halmos in his article, "The heart of mathematics," which appeared in the August-September, 1980 issue of the American Mathematical Monthly. The problem read as follows: Does there exist a disc (meaning interior and boundary of a circle) in the plane that contains exactly 71 lattice points? It so happens that the answer to this is yes. Figure 3-5 shows a sketch of a disc that could very well contain exactly 71 lattice points. The disc D is described by the set of all points that are at most r units away from some fixed center point (p,q): algebraically this would be stated,

$$D = \left\{ (x,y) \mid (x-p)^2 + (y-q)^2 \leq r^2 \right\}.$$

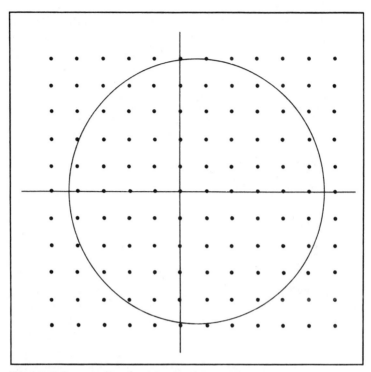

Fig. 3-5. A disc possibly containing 71 lattice points.

Suppose you wanted to find a disc that contains exactly 71 lattice points. This means you would need to find the appropriate center point (p,q), and the appropriate radius r. One thing that should be apparent is that the disc is not unique: there are in fact infinitely many such discs. For if (p,q) and r accurately detail one such disc, then (p+i,q+j) and r give another whenever i and j are integers. The presence of i and j serve merely to move the disc i units horizontally and j units vertically. On the other hand, a change in r, even a slight change, could significantly alter the number of lattice points in the disc. It may not alter it at all, which would happen if the boundary contained no lattice points and r is decreased by an amount less than the distance from the circle $(x-p)^2 + (y-q)^2 = r^2$ to the nearest interior lattice point. An increase in r would not alter the number of lattice points either as long as this increase is less than the distance between the circle and the nearest exterior lattice point.

For the sake of simplicity, suppose the center (p,q) of our disc is at the origin (0,0). Choosing various values of r; say r = 1/2, 2/2, 3/2, 4/2, ••• produces discs with increasing numbers of lattice points. Some of this data is given in Table 3-2.

Table 3-2. Various Discs and Their Lattice Points.

(p,q)	r	**Number of Lattice Points**
(0,0)	1/2	1
(0,0)	2/2	5
(0,0)	3/2	9
(0,0)	4/2	13
(0,0)	5/2	21
(0,0)	6/2	29
(0,0)	7/2	37

It appears that a pattern may be developing for the number of lattice points in the disc of radius r = n/2. But, aside from that, suppose you move the center (p,q) to a different location, and then complete a similar table. It will do no good to move (p,q) to another lattice point, since you will get exactly the same data. What you really need to do is move (p,q) around within a unit square; say within $0 \le p \le 1$, $0 \le q \le 1$. If you consider this unit square, as pictured in Fig. 3-6, you note that all eight points, P_1, P_2, •••, P_8, are located in a symmetric fashion in the unit square. Furthermore, any disc centered at P_i with radius r will contain the same number (certainly not the same ones) of lattice points as the disc centered at P_j with the same radius r. This means that if you want to move (p,q) around to truly different positions, it suffices to let (p,q) range through a half-portion of one of the quadrants of the square; say within $0 \le p \le 1/2$, $0 \le q \le p$. See Fig. 3-7.

To illustrate, if you first position (p,q) at (1/4,1/8) and then at (1/2,1/2), and varying r among 1/2, 2/2, 3/2, 4/2, •••, you get the results shown in Table 3-3. There are some noticeable differences in the number of lattice points contained within discs of equal radius, in particular when r = 3/2 or r = 5/2.

What you would like to do is write a program that directs the computer to search for a disc that contains exactly N lattice points. Halmos mentions, in his article, that such a disc exists for every positive integral value of N; an interesting result!

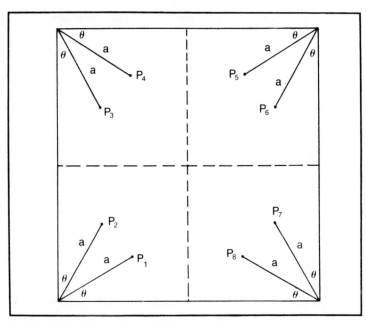

Fig. 3-6. Symmetry in the unit square.

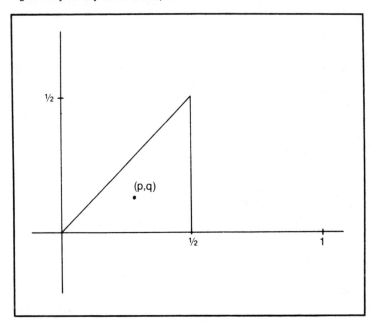

Fig. 3-7. A symmetric portion of the unit square.

Table 3-3. Lattice Points for Discs Centered Away from the Origin.

(p,q)	r	Number of Lattice Points
(1/4,1/8)	1/2	1
(1/4,1/8)	2/2	3
(1/4,1/8)	3/2	7
(1/4,1/8)	4/2	12
(1/4,1/8)	5/2	20
(1/4,1/8)	6/2	29
(1/2,1/2)	1/2	0
(1/2,1/2)	2/2	4
(1/2,1/2)	3/2	4
(1/2,1/2)	4/2	12
(1/2,1/2)	5/2	16
(1/2,1/2)	6/2	32

One part of the program will consist of varying the location of the center (p,q) of the disc in accordance with the region depicted in Fig. 3-7. Initially you may wish to divide the interval $[0,1/2]$ into an equal number of parts; refer to that number as PARTS. The distance between successive subdivisions becomes $1/(2*PARTS)$; this is shown in Fig. 3-8. If it turns out that none of these locations give exactly N lattice points (NLATTICE $\neq$ N), you can simply increase PARTS by, for example, 5, and run through the loop again. These statements in BASIC might be written as follows.

Program

```
100   INPUT PARTS, N
110   FOR I = 0 TO PARTS
120       P = I/(2*PARTS)
130       FOR J = 0 TO I
140           Q = J/(2*PARTS)
  •
  •
  •
200   REM COMPUTE NLATTICE POINTS
  •
  •
  •
400   PARTS = PARTS + 5
```

Assume that you have positioned (p,q). The second part of the program might consist of varying the radius r. For each value of r, you will need to tally the number of lattice points in the disc. Recall that any circle of radius r can be inscribed in a square of dimensions 2r by 2r. Hence, the square should contain at most approximately $4r^2$ lattice points. Furthermore, inside the circle, there should be at least $(r-1) + (r-1)$ lattice points on the horizontal diameter. Thus the number N of lattice points should be at least $[2(r-1)]^2$. In other words, for a given disc with center (p,q) and radius r, you should expect

$$4(r-1)^2 \leqslant N \leqslant 4r^2.$$

This inequality can be used to determine limits of r when a specific value of N is given. To search for a disc that contains 71 lattice points, for example, it would be appropriate to choose sample radii that range in value from $\sqrt{71/4} = 4.2$ to $\sqrt{71/4} + 1 = 5.2$. In this particular case, you may wish to increment r in steps of 0.1, or 0.2. For each value of r, the number (NLATTICE) of lattice points in the disc would be counted. Examine how this count might be made. Any lattice point (XLATT,YLATT) in the disc must satisfy the criteria

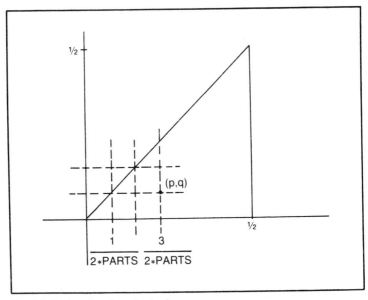

Fig. 3-8. Various locations for (p,q).

$$(XLATT - p)^2 + (YLATT - q)^2 \le r^2.$$

The point itself must be of the form

$$XLATT = [p - r + i] \qquad \text{with } i \in \{0,1,2,\cdots,2r\}$$
$$YLATT = [q - r + j] \qquad \text{with } j \in \{0,1,2,\cdots,2r\}$$

This is depicted in Fig. 3-9. The brackets around $p - r + i$ and $q - r + j$ are the mathematical symbols that indicate the greater integer

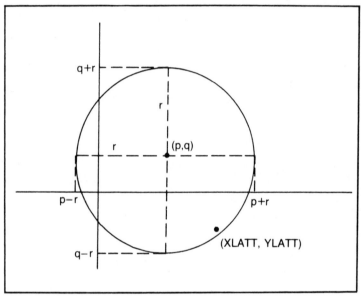

Fig. 3-9. Locating lattice points.

function. For the language BASIC, these bracket symbols are replaced by the function INT, which stands for integer. Some of the BASIC statements that could be a part of your program to search for the lattice points are listed below.

Program

```
205   NLATTICE = 0
210   FOR I = 0 TO 2*R
220       XLATT = INT(P - R + I)
230       FOR J = 0 TO 2*R
240           YLATT = INT(Q - R + J)
```

```
250              IF (XLATT − P)**2 + (YLATT − Q)**2
                 > R**2 GO TO 270
260                 NLATTICE = NLATTICE + 1
270        NEXT J
280   NEXT I
290   IF NLATTICE = N PRINT OUT ANSWER
300   REM   NOW TO CHANGE THE RADIUS
```

By now you have a sound understanding of the problem and some ideas (though they don't have to coincide with the ones discussed) on how to attack the problem. It is an interesting problem that combines features from both number theory and geometry, and forces you to construct efficient search algorithms. You may be interested to know that for each value of n, it is possible to find a circle of area = n that contains exactly n lattice points in its interior. Furthermore, you can find a circle that contains exactly n lattice points on its circumference [37].

Exercises

1. Write a program that locates a disc with a given number N of lattice points. Test your program with N = 71. Print out the value of the center (p,q), radius r, and the coordinates of the 71 lattice points.
2. What is the area of the polygon shown in Fig. 3-10?

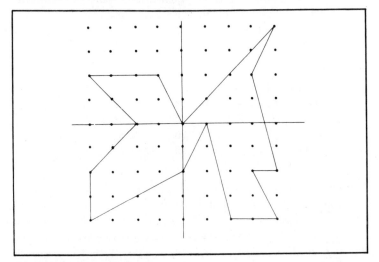

Fig. 3-10. A simple lattice polygon.

PRIMES

The prime numbers furnish us with as much challenging material as any subject matter could. These marvelous natural numbers, which are divisible evenly only by one and the number itself, include, 2, 3, 5, 7, 11, •••. It is customary to exclude 1 from the list, thereby making 2 the smallest (and the only even) prime. In this section some of the known results (the interesting ones) will be discussed; some conjectures that seem plausible yet are still unproven will be mentioned; and finally a result that furnishes a programmer with an elementary problem will be presented.

Although the notion of a prime is a very natural and obvious one, questions concerning the primes are often very difficult and many such questions remain unanswerable today. First, one rather crude but effective method for determining all the primes that are less than or equal to a given number N is known as the *Sieve of Eratosthenes*. Eratosthenes was a Greek mathematician (around 200 B.C.) who developed the sieve method. His other mathematical accomplishments are not nearly as well known as his method for determining primes. Many fail to realize that Eratosthenes was one of the first to argue that the earth was spherical, and he computed the diameter of the earth to within 50 miles of today's accepted value [57].

To use the sieve method, suppose, for example, you wish to find all the primes less than or equal to 50. You write down the numbers from 2 to 50. You then circle 2 (the first prime) and strike out all the succeeding multiples of 2 (4, 6, 8, •••, 48, 50). This gives the following:

	②	3	4̸	5	6̸	7	8̸	9	1̸0̸
11	1̸2̸	13	1̸4̸	15	1̸6̸	17	1̸8̸	19	2̸0̸
21	2̸2̸	23	2̸4̸	25	2̸6̸	27	2̸8̸	29	3̸0̸
31	3̸2̸	33	3̸4̸	35	3̸6̸	37	3̸8̸	39	4̸0̸
41	4̸2̸	43	4̸4̸	45	4̸6̸	47	4̸8̸	49	5̸0̸

You then circle the next unmarked number, 3 (the second prime), and strike out all multiples of 3 that are greater than 3 (some will have already been struck out). Then, you circle the next unmarked number, 5 (our third prime), and repeat the process. When you have finished, all the prime numbers less than or equal to 50 will be circled. See Fig. 3-11.

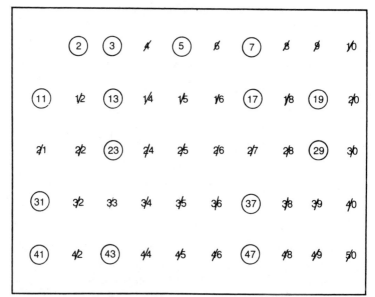

Fig. 3-11. Sieve of Eratosthenes.

It was Euclid who proved (Book IX, Prop. 20) that the set of all primes is infinite. His proof is one of the classics in all of mathematics, and for that reason I am presenting it here. Suppose $p_1 < p_2 < p_3 < \cdots < p_n$ are the only primes: just finitely many. If you form the number $N = p_1 p_2 \cdots p_n + 1$, it follows that when N is divided by each p_i, it leaves a remainder of 1. This means N has no proper divisors, so N must be prime. But N is greater than the largest prime p_n; so p_n must not have been the largest prime. This gives us a contradiction, thus establishing that the set of primes must be infinite.

This leads to an interesting conjecture, known as Fortune's conjecture (named after the anthropologist Reo Fortune, former husband to Margaret Mead), which first appeared in print in 1980 [34]. One question is, for which n is $N = p_1 p_2 \cdots p_n + 1$ prime? At present the only known prime values for N occur when n = 1,2,3,4,5 and 11. A small table of values for N is presented in Table 3-4. It is rather strange that so few of these values for N are prime. But if you let Q_n denote the smallest prime strictly greater than the associated value of N, so $Q_1 = 5$, $Q_2 = 11$, $Q_3 = 37$, $Q_4 = 223$, $Q_5 = 2333$, $Q_6 = 30047$, $Q_7 = 510529$, and $Q_8 = 9699713$, the numbers F_n formed by the difference $Q_n - (p_1 p_2 \cdots p_n)$ appear to always be prime! This sequence F_n of "fortunate numbers" has, for its first 20 terms,

Table 3-4. Various Values for $N = p_1 p_2 \cdots p_n + 1$.

n	N
1	$2 + 1 = 3$
2	$2 \cdot 3 + 1 = 7$
3	$2 \cdot 3 \cdot 5 + 1 = 31$
4	$2 \cdot 3 \cdot 5 \cdot 7 + 1 = 211$
5	$2 \cdot 3 \cdot 5 \cdot 7 \cdot 11 + 1 = 2311$
6	$2 \cdot 3 \cdot 5 \cdot 7 \cdot 11 \cdot 13 + 1 = 30031$
7	$2 \cdot 3 \cdot 5 \cdot 7 \cdot 11 \cdot 13 \cdot 17 + 1 = 510511$
8	$2 \cdot 3 \cdot 5 \cdot 7 \cdot 11 \cdot 13 \cdot 17 \cdot 19 + 1 = 9699691$

the values 3, 5, 7, 13, 23, 17, 19, 23, 37, 61, 67, 61, 71, 47, 107, 59, 61, 109, 89, 103. It is not known for sure whether all values of F_n are prime, but Fortune conjectured that they are, and every indication points to this being the case.

A number that is neither 1 nor a prime is said to be composite. Such a number can be represented as the product of two numbers, each greater than one. It is well known that by continued factorization you can eventually express any composite number as a product of primes, some of which may be repeated. Thus, for example, the number 300 can be expressed as the product

$$300 = 2 \times 2 \times 3 \times 5 \times 5$$

which can be more conveniently expressed as

$$300 = 2^2 \cdot 3 \cdot 5^2.$$

It is the Fundamental Theorem of Arithmetic that any composite number N (actually any natural number greater than one) can be expressed as the product of primes,

$$N = p_1^{a_1} p_2^{a_2} \cdots p_k^{a_k}$$

where the exponents a_i are integral, and the factorization is unique except for the ordering. The first clear statement and proof of this theorem seems to have been given by Gauss in his famous *Dis-*

quisitiones Arithmeticae of 1801. This fundamental theorem exhibits the structure of the natural numbers in relation to the operation of multiplication. It shows that the primes are the elements out of which all the natural numbers can be built by using multiplication in every possible way.

One important ramification of being able to express a number N as a product of primes is that it allows you to determine what and how many divisors N has. For, if N is expressed as

$$N = p_1{}^{a_1}p_2{}^{a_2}\cdots p_k{}^{a_k}$$

each divisor d is of the form

$$d = p_1{}^{b_1}p_2{}^{b_2}\cdots p_k{}^{b_k}$$

with each b_i satisfying $0 \leq b_i \leq a_i$. There are $(a_i + 1)$ choices for each b_i, so the total number of divisors has to be

$$(a_1 + 1)(a_2 + 1) \cdots (a_k + 1).$$

In the case with N = 300, since

$$300 = 2^2 \cdot 3^1 \cdot 5^2$$

you know there are $(2 + 1)(1 + 1)(2 + 1) = 18$ divisors of 300. These are listed in Table 3-5. If you would like to work on an interesting problem, it is still an open question (due to Erdos) whether there are infinitely many pairs of consecutive integers (n, n+1) where n and n+1 have the same number of divisors [23].

Except for 2, all the primes are odd. These odd primes can be conveniently separated into two classes, (a) and (b), as shown.

(a) 5, 13, 17, 29, 37, 41, 53, •••

(b) 3, 7, 11, 19, 23, 31, 43, 47, •••

The numbers in (a) are of the form 4n + 1, where n is some whole number, and those in (b) are of the form 4n + 3. One of these two forms must produce infinitely many primes, since the totality of primes is infinite. As it turns out, both classes produce infinitely

Table 3-5. Divisors of N = 300.

Divisors	Factorization	Divisors	Factorization
1	$2^0 3^0 5^0$	20	$2^2 3^0 5^1$
2	$2^1 3^0 5^0$	25	$2^0 3^0 5^2$
3	$2^0 3^1 5^0$	30	$2^1 3^1 5^1$
4	$2^2 3^0 5^1$	50	$2^1 3^0 5^2$
5	$2^0 3^0 5^1$	60	$2^2 3^1 5^1$
6	$2^1 3^1 5^0$	75	$2^0 3^1 5^2$
10	$2^1 3^0 5^1$	100	$2^2 3^0 5^2$
12	$2^2 3^1 5^0$	150	$2^1 3^1 5^2$
15	$2^0 3^1 5^1$	300	$2^2 3^1 5^2$

many primes. It is also a fact that there are infinitely many primes of the form $6n + 5$ and $8n + 5$. All these results are special cases of a famous theorem of Dirichlet [36].

Theorem: If a and b are natural numbers with no common divisor except 1, there are infinitely many primes of the form $an + b$.

There is a conjecture that there are infinitely many primes of the quadratic form $n^2 + 1$. The primes 2, 5, 17, 37, 101, and 197 fall into this category. This is as of yet proven, but recent work [28] tends to support the claim. For each prime p, there appears to be a smallest integer k for which $4p^2k^2 + 1$ is a prime. This has been checked by a Hewlett Packard computer for all primes less than 5000. In all cases, the largest value of k was 45.

In 1742 the mathematician Christian Goldbach conjectured that every even number greater than 4 could be expressed as the sum of two odd primes. Thus you can write $6 = 3 + 3$, $8 = 3 + 5$, $10 = 5 + 5$, $12 = 5 + 7$, $14 = 7 + 7$, $16 = 7 + 9$, and so on. What a marvelously simple conjecture: anyone can understand it, but no one has been able to prove or disprove it. Likewise consider Goldbach's second conjecture that every odd number greater than 7 can be expressed as the sum of three odd primes. You may wish to give some thought to writing a program that determines the three odd primes for each odd number greater than 7.

In an attempt to discover laws governing those expressions that generate prime numbers, Mersenne had thought that numbers of the form $2^p - 1$, for a prime p, would produce a prime. Mersenne declared that $2^p - 1$ was prime for the specific values p = 2, 3, 5, 7, 13, 17, 19, 31, 67, 127 and 257, and composite for all other values of p smaller than 257. Several flaws were found in this. For instance, p = 257 and p = 67 did not yield a prime, while p = 61 and p = 107 did. Meanwhile, Fermat suggested that the expression

$$2^{(2^n)} + 1$$

would generate only prime numbers for non-negative integral values of n. For instance, n = 0, 1, 2, 3, 4 produce the (Fermat) numbers 3, 5, 17, 257, 65537, which are all primes. But Euler showed that for n = 5, the Fermat number 4,294,967,297 is composite (641 × 6,700,417). Modern day computers have identified the next 34 Fermat numbers and shown them all to be composite.

In 1772 Euler showed that the expression $n^2 + n + 41$ produces a prime for each n = 0, 1, 2, •••, 39—quite a remarkable consecutive string of primes. The expression $n^2 - 79n + 1601$ produces a prime (though not all different) for the 80 consecutive integral values n = 0, 1, 2, •••, 79. These are quite productive quadratics. Higher degree polynomial expressions that produce more primes can be found. However, no polynomial except those of degree one, an + b, are known to yield any more than finitely many different primes. There are polynomial expressions of two variables that yield infinitely many prime values. One such expression is $n^2 + m^2 + 1$. But more remarkable is the expression

$$\frac{(m-1)}{2} \left[|B^2 - 1| - (B^2 - 1) \right] + 2$$

where B = n(m + 1) - (m! + 1), which yields only primes, every possible prime, and each odd prime exactly once [38].

Now look at one of the more fascinating and mysterious functions in mathematics. For each number x, let $\pi(x)$ represent the total number of primes p with p ≤ x. Table 3-6 lists some of the values of this function.

Table 3-6. Values of $\pi(x)$.

x	$\pi(x)$	x	$\pi(x)$
1	0	10	4
2	1	100	25
3	2	1000	168
4	2	10,000	1229
5	3	100,000	9592
6	3	1,000,000	78,498
7	4	10,000.000	664,579
8	4	100,000,000	5,761,455
9	4	1,000,000,000	50,847,534

The graph of this function $y = \pi(x)$, which is a nondecreasing function, would appear as an infinite stair-step graph. The graph would exhibit a unit jump with every occurrence of a prime x, and the graph would remain level for each composite. Mathematicians desire to predict with some certainty whether the graph at the particular value or range of values would increase or remain level. This is virtually impossible to predict, and that is one reason people continue to study $\pi(x)$. Each one who engages in this study hopes to discover some secret law governing the behavior of the distribution of primes.

Several amazing results concerning the distribution are mentioned in [36], [39] and [79]. The famous prime number theorem (Gauss discovered it at the age of fifteen) asserts that $\pi(x)$ is approximately equal to x/ln(x),

$$\pi(x) \doteq \frac{x}{\ln(x)}$$

and the approximation improves as x approaches infinity. We see from Table 3-6, that if x = 1,000,000,000, $\pi(x) = 50,847,534$ and x/ln(x) = 48,254,942. An integral approximation, denoted by Li(x), known as the logarithmic integral and defined by

$$Li(x) = \int_{2}^{x} 1/(\ln(t))\ dt$$

serves as an even better approximation to $\pi(x)$. For all realistic values of x, the values of $\pi(x)$ are always less than Li(x), and no numbers x have been found that reverse this inequality. Littlewood proved, however, that numbers x exist for which Li(x) $< \pi(x)$, but was unable to exhibit any, and was also unable to determine any upper bound for such x. It was left to S. Skewes, a student of Littlewood, to prove that there is at least one value x with Li(x) $<$ $\pi(x)$, and with x less than the number

$$10^{10^{10^{34}}}$$

which is known as Skewes' number. Quite a bound!

Suppose it is known that x $=$ p is a prime. Where does the next prime occur? This is a most difficult question to answer. A partial answer is given by Bertrand's Postulate, which asserts the existence of a prime between any integer n and 2n. But as often as primes occur, there can be arbitrarily long stretches of composite numbers. The reader may wish to find a stretch of, say, 20 consecutive composite numbers. An interesting result, discovered by the Polish mathematician Sierpinski [39], asserts that for every number n, there exists a prime p such that the n natural numbers on each side of it (i.e., p+1, p+2, •••, p+n and p−1, p−2, •••, p−n) are all composite numbers! Again, as soon as you start thinking that there aren't really that many primes in comparison to composites or that their density is not too great, it should be pointed out that the primes do occur with high enough frequency to cause the series of

$$\sum_p 1/p$$

reciprocals of primes to diverge [55].

Finally, let us discuss a property that would interest (many of the aforementioned results should already be of interest) the programmer. Each natural number n can be decomposed into a product of primes

$$n = p_1^{a_1} p_2^{a_2} \cdots p_k^{a_k}$$

as asserted by the Fundamental Theorem of Arithmetic. In this form you can see that n has k different prime factors. This leads to defining a function, f(n), which represents the number of different prime factors of n. For example, since n = 300 has the decomposition

$$300 = 2^2 \cdot 3^1 \cdot 5^2$$

with three different prime factors, it follows that f(300) = 3. Likewise, some other values of f include f(12) = 2, f(27) = 1, f(990) = 4. The function f behaves quite irregularly because the factors of one integer are completely different from the factors of the next integer. There is no correlation to the quantity of factors possessed by consecutive integers or, for that matter, by two integers chosen at random. On the other hand, the average value of f behaves a lot more regularly. To be more explicit, the average,

$$\frac{f(2) + f(3) + \cdots + f(n)}{n}$$

is asymptotic to ln(ln(n)), and thus the approximation improves as n increases. It's ironic how often the function ln(ln(n)), or ln(ln(ln(n))) appears in connection with a prime number search. You may wish to read [44], which presents an interesting discussion of the computer time needed to determine whether a given number (a very large number) is prime, because time is asymptotic to a composition of logarithm functions.

According to [36], the average of f can be written more precisely as

$$(1/n) \sum_{i = 2}^{n} f(i) \doteq \ln(\ln(n)) + B_1$$

where B_1 is the expression

$$B_1 = \gamma + \sum_{\substack{p \\ p \leq n}} [\ln(1 - 1/p) + 1/p]$$

In this expression, γ is Euler's constant ($\gamma = .523$), and the sum extends over all the primes that do not exceed n. For the sake of simplicity this approximation can be rewritten as

78

Table 3-7. Values and Approximates for Σf(i).

n	$\sum_{i=2}^{n} f(i)$	RHS
5	4	3.55
10	11	10.57
15	19	18.18
20	26	26.19
•		
•		
•		
980	2081	2094.32
985	2091	2105.73
990	2106	2117.15
995	2116	2128.57
1000	2126	2139.99

$$\sum_{i=2}^{n} f(i) \doteq n\ln(\ln(n)) + n\gamma + n \sum_{\substack{p \\ p \geq n}} [\ln(1 - 1/p) + 1/p]$$

If you denote the right-hand side of this approximation by R.H.S., some sample data is shown in Table 3-7.

You wish to write a program that extends this table, say for n up to some upper limit UPPER. Now what does this involve? Obviously you will have to know the primes $p_1 = 2, p_2 = 3, p_3 = 5, \cdots, p_k$ where this last prime p_k will be approximately the square root of UPPER. One can either program the computation of these primes or have them fed in as input. For each value of i, the number of prime factors of i will be f(i). To determine which factors these are, you only need to test for divisibility those primes $p = \sqrt{i}$. A running total of values of f(i) will be compared with the appropriate value of R.H.S. from above.

Exercises

1. Write a program that computes the values of f(n) for n = n = 2, 3, $\cdots$, UPPER where UPPER is some prescribed integer. Compare the running total f(2) + f(3) + $\cdots$ + f(n) with the expression

$$n\ln(\ln(n)) + n\gamma + n \sum_{\substack{p \\ p \leq n}} [\ln(1 - 1/p) + 1/p].$$

Print out your results for n = 2, 3, •••, 50 and for n = UPPER − 4, UPPER − 3, UPPER − 2, UPPER − 1, UPPER.

2. If Skewes' numbers were to be multipled out, how many digits would it contain?

3. The largest known Mersenne prime is M_{44497}, or
$$2^{44497} - 1.$$

Use logarithms to determine how many digits this prime has.

WEIRD NUMBERS

I don't think there is any doubt that the set of prime numbers ranks high on the list of the importance of all subsets of integers with regard to the quantity of research literature devoted to it. This is deservedly so because not only can the primes be viewed as the building blocks of the natural numbers and hence of the real number system, but the primes also possess so many intriguing, mysterious, and beautiful results that one is continually overwhelmed and amazed at their role in number theory. I intend to present some classifications of numbers of lesser renown. Certainly most of these numbers will be of minimal importance and will be of interest only to those few who enjoy the aesthetic beauty of studying number related properties. As is quite often the case, these properties furnish challenging exercises for the programmer.

I begin by letting N denote the set of natural numbers, which include 1, 2, 3, •••. Typically the symbols n and m will denote some element of N. Many functions exist on N, but the only one I am interested in is the "sum of the divisors" function. This function, denoted by θ, is defined by $\theta(n)$ equaling the sum of all the natural number divisors of n;

$$\theta(n) = \sum_{m|n} m.$$

To illustrate, values for $\theta(6)$, $\theta(12)$, $\theta(30)$, $\theta(35)$ and $\theta(41)$ are worked out in Table 3-8. It is obvious that every natural number n > 1 has at least two divisors, namely 1 and n. Hence $\theta(n)$ is at least equal to n + 1: this would be the exact value of $\theta(n)$ whenever n is a prime. Early in our history it was considered important to know whether, and when, the proper divisors of n (all the divisors except n itself) summed to n. Alternatively, does $\theta(n) = 2n$, or is $\theta(n) < 2n$ or $\theta(n) > 2n$? These cases are mutually exclusive, and form a natural subdivision of N into three classes. More specifically,

Table 3-8. Some Values of $\theta(n)$.

n	Divisors of n	$\theta(n)$
6	1,2,3,6	12
12	1,2,3,4,6,12	28
30	1,2,3,5,6,10,15,30	72
35	1,5,7,35	48
41	1,41	42

a natural number n is defined to be deficient, perfect, or abundant depending on whether $\theta(n)$ is less than, equal to, or greater than 2n, respectively. The Venn diagram in Fig. 3-12 depicts the classification.

From the previous table you can see that 35 and 41 are examples of deficient numbers, that 6 is a perfect number, and 12 and 30 are abundant numbers. These distinctions were considered impor-

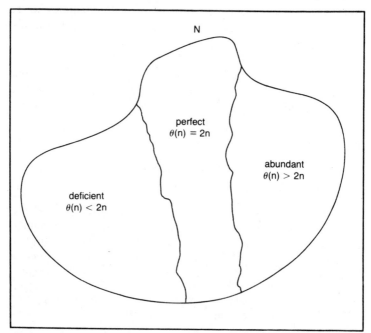

Fig. 3-12. A natural subdivision of the natural numbers.

tant in ancient numerology. Perfect numbers were considered the most aesthetically pleasing. They were to represent the perfect medium between abundance and deficiency and were the essential elements in all numerological speculations. God created the world in 6 days and the moon circles the earth in 28 days: two examples of perfection in the best of all possible worlds [60].

In classifying the natural numbers into these three categories, several properties can easily be established.

Property 1: Every prime p is a deficient number.

Property 2: Every power of a prime, p^n, is deficient.

Property 3: Every divisor of a deficient number is deficient.

Property 4: Every proper divisor of a perfect number is deficient.

Property 5: If n is a perfect number, then each multiple mn (m ≥ 2) is abundant.

Property 6: If n is any abundant number, each multiple mn is abundant.

To see why these last two properties are true, let n be a perfect or abundant number, so $\theta(n) \geq 2n$. The numbers $d_1, d_2, \cdots, d_k$ denote the divisors of n. This gives

$$\theta(n) = \sum_{i=1}^{k} d_i \geq 2n.$$

Choose an arbitrary natural number m ≥ 2, and the claim is that mn is abundant. To this end, note that some (not all) of the divisors of mn are $m, md_1, md_2, \cdots, md_k$. Summing these divisors gives a sum greater than 2mn,

$$\theta(mn) > m + md_1 + md_2 + \cdots + md_k = m + m \sum_{i=1}^{k} d_i$$
$$\geq m + 2mn$$
$$> 2mn$$

thus implying mn is abundant.

Because the deficient numbers and abundant numbers are so plentiful, the majority of the literature is devoted to perfect numbers. These numbers are very rare, with only several dozen known. The first four are 6, 28, 496, and 8128. Clearly the perfect numbers grow at a rapid rate. Euclid showed that any number of the form $2^{n-1}(2^n - 1)$ is perfect, providing the factor $2^n - 1$ is prime. For instance, when $n = 2$ the factor $2^n - 1 = 3$ is prime, which produces the perfect number 6. When $n = 3$, the perfect number 28 is produced, and for $n = 5$ the result is 496. Euler established the converse of this representation, namely that if a perfect number was even, it must be factorable as $2^{n-1}(2^n - 1)$, with the odd factor being prime. As for odd perfect numbers, none have been found yet, and it is unknown whether one exists. By the use of modern calculators [71] it has been proven that if an odd perfect exists, it must be extremely large; it must be greater than 10^{36}.

In recent times the notions of deficient, perfect, and abundant have been generalized to include such numbers as quasiperfect, semiperfect, multiply perfect, superperfect, superduperperfect, practical, amicable and superabundant [37]. These definitions are all similar in nature, and all depend on the function θ.

To begin with, the function θ^* is defined by $\theta^*(n)$, which is the sum of all the proper divisors of n. Thus, for $n \geq 2$, you have

$$\theta^*(n) = \sum_{\substack{d|n \\ d \neq n}} d = \theta(n) - n.$$

A number n is, therefore, perfect if $n = \theta^*(n)$. The number n is defined to be superperfect if $n = \theta^*(\theta^*(n))$, and superduperperfect if $n = \theta^*(\theta^*(\theta^*(n)))$. The results from Table 3-9 show $n = 6$ to be both superperfect and superduperperfect.

A number n is said to be multiply-perfect if the sum of all its divisors is kn, where k is some natural number. Such numbers n are said to be k-tuply perfect. The case $k = 2$ corresponds to perfect numbers. The number 120 is triply-perfect because $\theta(120) = 360 = 3(120)$, while $n = 2178540$ is 4-tuply perfect. Some results are known concerning multiply perfect numbers. It is known that if n is a triply-perfect number that is not divisible by 3, then 3n is a 4-tuply perfect number; and that every 5-tuply perfect number must have more than 5 different prime factors. But several important questions still remain unanswered. Such as, how many k-tuply numbers are there? Are there infinitely many? Are all k-tuply perfect numbers even? You may wish to pursue some investigation into these matters.

Table 3-9. Some Values of $\theta^*(n)$, $\theta^*(\theta^*(n))$, and $\theta^*(\theta^*(\theta^*(n)))$.

n	$\theta^*(n)$	$\theta^*(\theta(n))$	$\theta^*(\theta^*(\theta^*(n)))$
2	1	—	—
3	1	—	—
4	3	1	—
5	1	—	—
6	6	6	6
7	1	—	—
8	7	1	—
9	4	3	1
10	8	7	1
11	1	—	—
12	16	15	9
13	1	—	—
14	10	8	7
15	9	4	3
16	15	9	4
17	1	—	—
18	21	11	1
19	1	—	—
20	22	14	10

Suppose you consider the sequence of ratios $\theta(n)/n$ as shown in Table 3-10. These values tend to oscillate a lot. With the sole exception of $n = 1$, all the values of $\theta(n)/n$ are greater than one. There is a subsequence with values that converge to one, namely those terms when $n = p$ is prime,

$$\lim_{p \to \infty} \theta(p)/p = 1.$$

A number n is therefore deficient if $\theta(n)/n < 2$, and abundant if $\theta(n)/n > 2$. If $\theta(n)/n$ is greater than $\theta(m)/m$ for all $m < n$, n is defined to be superabundant. It follows from the table that n = 2, 4, 6, and 12 are superabundant. Note that 2 and 4 are deficient, and are the only deficient numbers that are superabundant because any superabundant number greater than 12 must have a ratio $\theta(n)/n$ greater than 2.33••• , which means it is abundant. But which abundant numbers greater than 12 have a ratio in excess of 2.3••• ? There are, in fact, infinitely many. More surprising is that there exists an infinite sequence $\{n_i\}$ of natural numbers, all abundant, for which $n_1 < n_2 < n_3 < \bullet\bullet\bullet$ and

$$\frac{\theta(n_i)}{n_i} < \frac{\theta(n_{i+1})}{n_{i+1}}$$

with $\lim \theta(n_i)/n_i = \infty$. This remarkable result establishes an infinitude of superabundant numbers. You should wonder whether there is a good way to characterize these numbers.

Somewhat less interesting is the notion of quasiperfect numbers. Here, n is termed quasiperfect if it is almost equal to the sum of its proper divisors. If n is anything other than perfect, the closest it could come to equaling the sum of its proper divisors, $\theta^*(n)$, is to differ from it by ± 1. Thus, n is quasiperfect if $n = \theta^*(n) \pm 1$. No

Table 3-10. Values of $\theta(n)/n$.

n	$\theta(n)$	$\theta(n)/n$
1	1	1
2	3	1.5
3	4	1.3333
4	7	1.75
5	6	1.2
6	12	2.
7	8	1.1428
8	15	1.875
9	13	1.4444
10	18	1.8
11	12	1.0909
12	28	2.3333
13	14	1.0769
14	24	1.7142
15	24	1.6
16	31	1.9375
17	18	1.0588
18	39	2.1666
19	20	1.0526
20	42	2.1

solutions are known to the equation $n = \theta^*(n) - 1$ (these would be abundant numbers that are minimally abundant), while all powers of 2 do satisfy $n = \theta^*(n) + 1$ (these are deficient numbers that are minimally deficient). You may wish to study the solutions to $n = \theta^*(n) \pm k$ for each natural number k and call these numbers k-quasiperfect numbers.

Semiperfect numbers are those numbers n for which n is equal to the sum of some collection of proper divisors of n. A perfect number is equal to the sum of all its proper divisors, so it is semiperfect also. Some numbers can definitely not be expressed as the sum of some collection of proper divisors; deficient numbers are ruled out by definition. How about abundant numbers? The amazing thing is that practically every abundant number is semiperfect. Table 3-11 shows how the first ten abundant numbers can be so

Table 3-11. Representing Abundant Numbers as a Sum of Their Proper Divisors.

Abundant n	Proper Divisors of n	Sum representation
12	1,2,3,4,6	1 + 2 + 3 + 6
18	1,2,3,6,9	3 + 6 + 9
20	1,2,4,5,10	1 + 4 + 5 + 10
24	1,2,3,4,6,8,12	4 + 8 + 12
30	1,2,3,5,6,10,15	5 + 10 + 15
36	1,2,3,4,6,9,12,18	1 + 2 + 3 + 12 + 18
40	1,2,4,5,8,10,20	1 + 4 + 5 + 10 + 20
42	1,2,3,6,7,14,21	1 + 6 + 14 + 21
48	1,2,3,4,6,8,12,16,24	8 + 16 + 24
54	1,2,3,6,9,18,27	3 + 6 + 18 + 27

expressed as a sum of proper divisors. Abundant numbers possess so many different divisors (otherwise they wouldn't be abundant) that usually some combination of divisors will sum exactly to the number. The number 36, for example, has several different sets of divisors that sum to 36. One way would be $6 + 12 + 18$, while three others would be $1 + 2 + 3 + 12 + 18$. $3 + 6 + 9 + 18$, and $2 + 3 + 4 + 9 + 18$.

But, alas, not every abundant number has this property. Note that n = 70 is abundant because its divisors 1, 2, 5, 7, 10, 14, 35, 70 sum to more than 2(70). None of the proper divisors, however, sum to 70. This is our first example of an abundant nonsemiperfect number. In fact it is the only one found among the first 500 natural numbers. These examples are so scarce, they have been termed *weird numbers* [37]. It is known that there are infinitely many weird numbers, but they are difficult to characterize. The conjecture is that they are all even; this evenness property seems to be evident in many of these categories, as is the case with quasiperfect, k-tuply perfect, and perfect.

The second smallest weird number falls in the interval $500 < n < 1000$. You can program a search for it. This means that for each n in the interval, you will need to obtain the divisors of n; say, DIVIS(1), DIVIS(2), •••, DIVIS(k). You can arrange these in increasing order with $1 = $ DIVIS(1), $n = $ DIVIS(k), and DIVIS(i) < DIVIS(i + 1). You will want to make sure n is abundant, so the sum DIVIS(1) + DIVIS(2) + ••• + DIVIS(k) should be at least as great as 2n. Then you will need to find an ingenious and efficient way to search through the set of proper divisors for that subset which sums to n exactly. It is recommended that you study the divisors listed in Table 3-11 and try to determine a workable algorithm for obtaining the appropriate sum representation.

Exercises

1. Write a program that locates the weird number n located between 500 and 1000.
2. What is the smallest odd abundant number?

EUCLID'S ALGORITHM

Consider the following problem, which could be typical of many small business firms.

> The Huron Cycle Shop is to open its doors for the first time early this spring. It is in the process now of stocking its shelves with sufficient inventory. The manager has room to carry at most 80 bicycles, and so placed a total order of $2490 for some men's bicycles at $29 each and some women's bicycles at $33 each. How many bicycles of each did the manager order?

If you denote the number of men's bicycles that were ordered by m and the number of women's bicycles ordered by w, this problem reduces to solving the system

$$29m + 33w = 2490$$
$$m + w \leqslant 80.$$

Begin by focusing your attention on the linear equations of the form $ax + by = c$ where a, b, and c are given integers, and the desired solution integers x and y, are also integers. Quite often the solution integers must be positive or else the solution is without meaning. Such is the case with the Huron Cycle Shop problem because the only integral solutions that make sense are positive.

Any divisor d of both a and b must also divide $ax + by$, and hence must divide c. If not, the equation has no solution. Thus, the equation $12x + 6y = 8$ has no (integral) solution because $d = 3$ divides into both 12 and 6, but not 8. The equation $8x - 12y = 10$ also fails to have a solution because $d = 4$ divides into $8x - 12y$, but not 10. Even though the divisor $d = 2$ divides into both $8x - 12y$ and 10, this is not enough to guarantee that the equation is solvable. The equation $ax + by = c$ is solvable every time d divides c for every d that is a common divisor of both a and b. Since this is the case, you might as well deal with the largest such divisor d, since every other divisor must divide into the largest divisor. The notation $\gcd(a,b)$ will represent the greatest common divisor of both a and b. You have, for example, $\gcd(8,12) = 4$ and $\gcd(12,6) = 6$.

There will be a further discussion of this greatest common divisor $\gcd(a,b)$ later. Now let's go through the remaining analysis required to solve the linear equation $ax + by = c$. If x_0, y_0 is one solution that was discovered by some lucky means, then $ax_0 + by_0 = c$. But, knowing one integral solution allows you to generate infinitely many more integral solutions. For there is another solution given by x_1, y_1 with

$$x_1 = x_0 + \frac{b}{\gcd(a,b)}$$

$$y_1 = y_0 - \frac{a}{\gcd(a,b)}$$

because

$$ax_1 + by_1 = a \left[x_0 + \frac{b}{\gcd(a,b)} \right] + b \left[y_0 - \frac{a}{\gcd(a,b)} \right]$$

$$= ax_0 + by_0 + \frac{ab}{\gcd(a,b)} - \frac{ba}{\gcd(a,b)}$$

$$= c.$$

Likewise, $x_2 = x_0 + 2b/\gcd(a,b)$, $y_2 = y_0 - 2a/\gcd(a,b)$ is another solution; in general,

$$x_k = x_0 + \frac{bk}{\gcd(a,b)}$$

$$y_k = y_0 - \frac{ak}{\gcd(a,b)}$$

gives a solution for each integral value of k, $k = 0$, ± 1, ± 2, •••. To illustrate this result, consider the equation $4x - 6y = 18$, and note that a solution exists because $\gcd(4,6) = 2$ and 2 divides 18. A little searching gives the initial solution $x_0 = 3$, $y_0 = -1$. Hence the general solution is $x_k = 3 - 6k/2 = 3 - 3k$, $y_k = -1 - 4k/2 = -1 - 2k$. The data listed in Table 3-12 furnishes some more solutions to the equation $4x - 6y = 18$.

Table 3-12. Some Solutions (x_k, y_k) to $4x - 6y = 18$.

k	x_k	y_k
−2	9	3
−1	6	1
0	3	−1
1	0	−3
2	−3	−5

The graphical interpretation of this situation is that $4x - 6y = 18$, which graphs as a straight line (with slope = 4/6 y-intercept = 3) in the plane, will pass through infinitely many lattice points. The points $(3, -1)$, $(0, -3)$ and $(6, 1)$ are just three such points as shown in Fig. 3-13. In general, the graph of $ax + by = c$ will pass through either infinitely many lattice points or no lattice point, depending on whether $\gcd(a,b)$ divides c.

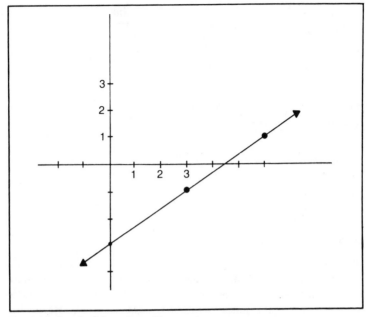

Fig. 3-13. Graph of $4x - 6y = 18$.

So, the question now becomes, how do you locate one solution, an initial solution x_0, y_0, to the equation $ax + by = c$? This is where a procedure known as the Euclidean algorithm comes into play. First, it is important to list a few properties of the operation of finding the greatest common divisor $\gcd(a,b)$ of two integers a and b.

Property 1: If a and b are factored completely (the fundamental theorem of arithmetic) so that

$$a = p_1^{\alpha_1} p_2^{\alpha_2} \bullet\bullet\bullet p_k^{\alpha_k}$$
$$b = p_1^{\beta_1} p_2^{\beta_2} \bullet\bullet\bullet p_k^{\beta_k}$$

then the greatest common divisor of a and b is given by

$$\gcd(a,b) = p_1^{\min(\alpha_1, \beta_1)} p_2^{\min(\alpha_2, \beta_2)} \bullet\bullet\bullet p_k^{\min(\alpha_k, \beta_k)}.$$

Property 2: If m is a positive integer, then

$$\gcd(ma, mb) = m \bullet \gcd(a,b).$$

Property 3: If m is a positive integer, and m divides both a and b, then

$$\gcd\left(\frac{a}{m}, \frac{b}{m}\right) = \frac{1}{m} \gcd(a,b).$$

Property 4: $\gcd(a,b) = \gcd(a,b-a) = \gcd(a-b,b)$.

Property 1 furnishes a useful means for computing the largest common divisor of two numbers if you are willing to compute the prime factorization of each. For instance, if $a = 36,300$ and $b = 43,510$ then their prime factorization is

$$36300 = 2^2 3^1 5^2 11^2$$
$$48510 = 2^1 3^2 5^1 7^2 11^1$$

so

$$\gcd(36300,48510) = 2^1 3^1 5^1 7^0 11^1 = 330.$$

To program this method, you would need to compute the factorization of each integer. This would be equivalent to determining the exponents $\alpha_1, \alpha_2, \cdots, \alpha_k$ on the successive primes. These could then be stored in an array. The prime factorization of $a = 36300$ may be stored as the array $A = (2,1,2,0,2,0,0,\cdots,0)$, while $b = 48510$ would be stored as $B = (1,2,1,2,1,0,0,\cdots,0)$. Comparison of $A(i)$ with $B(i)$ would give the greatest common divisor array $GCD(i)$ where

$$GCD(i) = \min\{A(i), B(i)\}.$$

The computer has to perform quite a few calculations to obtain the prime factorization of each integer, but it is a simple algorithmic procedure.

But direct your attention instead to property 4, which lays the foundation for Euclid's algorithm. Since $\gcd(a,b)$ is equal to both $\gcd(a,b-a)$ and $\gcd(a-b,b)$ it is obvious that one of $a-b$ and $b-a$ is positive while the other is negative, as long as $a \neq b$. Make it a point to always work with the difference that is positive. Using $a = 36,300$ and $b = 48,510$ you can compute $\gcd(a,b)$ by repeatedly using property 4 as follows:

$$
\begin{aligned}
\gcd(36300,48510) &= \gcd(36300,48510-36300) \\
&= \gcd(36300,12210) \\
&= \gcd(36300-12210,12210) \\
&= \gcd(24090,12210) \\
&= \gcd(11880,12210) \\
&= \gcd(11880,330) \\
&= \gcd(11550,330) \\
&\quad\bullet \\
&\quad\bullet \\
&\quad\bullet \\
&= \gcd(660,330) \\
&= \gcd(330,330) \\
&= 330.
\end{aligned}
$$

This procedure can be easily programmed as shown by the flowchart in Fig. 3-14. The integer listed as the first coordinate of the ordered pair will be denoted by FIRST and the second integer by SECOND. The greatest common divisor of a and b will be determined as soon as FIRST equals SECOND.

You might have noted during the previous computation of gcd(36300,48510) that the sequence of steps leading from gcd(11880,330) to gcd(330,330) actually consists of 35 steps—the number 330 is involved in the subtraction process a total of 35 times. It would be more efficient to perform the subtraction once, namely $1180 - 330(35)$. In order to do this you would have to have computed the number 35. How would you have done this? By dividing 11880 by 330, you get 36, which means 11880 is a perfect multiple of 330. This means that the algorithm could have been extended to

$$
\begin{aligned}
&= \gcd(11550,330) \\
&\quad\bullet \\
&\quad\bullet \\
&\quad\bullet \\
&= \gcd(660,330) \\
&= \gcd(330,330) \\
&= \gcd(0,330)
\end{aligned}
$$

so that the first coordinate is zero. If this happens (or if either coordinate is zero), the second coordinate represents the value of gcd(a,b). In general, given the situation of finding gcd(FIRST,SEC-

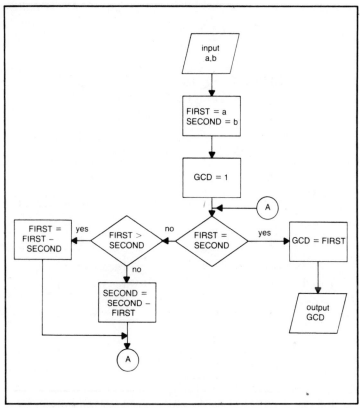

Fig. 3-14. Flowchart for determining greatest common divisor.

OND), you must divide the smaller number into the larger, and that number rounded off to the next smaller integer represents how many multiples of the smaller number need to be subtracted from the larger number. To illustrate, in order to evaluate gcd(472,36), you would divide 472/36 and get 13.11•••, which means

$$gcd(472,36) = gcd(472 - 36 \cdot 13,36)$$
$$= gcd(4,36).$$

This procedure allows you to alter the flowchart so it now reads as shown in Fig. 3-15.

This is the procedure commonly known as Euclid's algorithm, which is discussed in most texts on number theory. The way the method is usually stated is as follows.

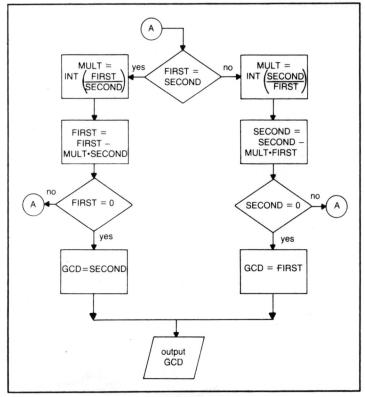

Fig. 3-15. Revised flowchart for computing greatest common divisor.

Euclid's Algorithm: Let a and b be two positive integers, and suppose $a \geq b$. There exists unique integers q_1 and r_1 (the quotient and the remainder) for which $a = bq_1 + r_1$, with $0 \leq r_1 < b$. If $r_1 = 0$, the number b is the greatest common divisor of a and b. If the remainder $r_1 \neq 0$, continue the pattern by obtaining a sequence of quotients $q_1, q_2, \bullet\bullet\bullet, q_{k+1}$ and a sequence of remainders $r_1, r_2, \bullet\bullet\bullet, r_k$ such that

$$a = bq_1 + r_1 \qquad 0 < r_1 < b$$
$$b = r_1q_2 + r_2 \qquad 0 < r_2 < r_1$$
$$r_1 = r_2q_3 + r_3 \qquad 0 < r_3 < r_2$$

$$\vdots$$

$$r_{k-2} = r_{k-1}q_k + r_k \qquad 0 < r_k < r_{k-1}$$
$$r_{k-1} = r_k q_{k+1}.$$

When you finally arrive at a zero remainder, $r_{k+1} = 0$, the previous remainder r_k proves to be the value of gcd(a,b).

The Euclidean algorithm is actually quite practicable in numerical cases; for example, if you wish to evaluate gcd(48510,36300), compute as follows:

$$48510 = 36300 \cdot 1 + 12210$$
$$36300 = 12210 \cdot 2 + 11880$$
$$12210 = 11880 \cdot 1 + 330$$
$$11880 = 330 \cdot 36.$$

Hence gcd(48510,36300) = 330.

You can work backwards from this string of equalities and get

$$330 = 12210 - 11800 \cdot 1$$
$$= 12210 - 1[36300 - 12210 \cdot 2]$$
$$= 12210 \cdot 3 - 1 \cdot 36300$$
$$= 3[48510 - 36300] - 1 \cdot 36300$$
$$= 3 \cdot 48510 - 4 \cdot 36300.$$

This furnishes a solution to the linear equation

$$48510x + 36300y = 330,$$

namely $x = 3$, $y = -4$. This means that you should be able to find a solution to the general equation $ax + by = gcd(a,b)$, or even more so, to the equation $ax + by = c$.

Exercises

1. Write a program that uses Euclid's algorithm to determine the greatest common divisor of two given positive integers a, b. Test your program with the following pairs of input data.

a) $a = 48510,$ $b = 36300$
b) $a = 10672,$ $b = 4147$
c) $a = 6552,$ $b = 100386$

2. How many bicycles of each kind did the Huron Cycle Shop order?

POWERFUL NUMBERS

Another of the "exotic" numbers to join the list that already includes the superabundant, superduperperfect, and practical numbers is that of powerful numbers. This generalization is defined by Golomb [33] as follows: a positive integer n is said to be powerful if p^2 divides n for every prime p that divides n.

The meaning behind the word *powerful* is vague, but apparently an integer is deemed nonpowerful if it lacks a second prime factor for some prime. It follows that any even powerful number must be divisible by 4, and any powerful number that is a multiple of 3 must be divisible by 9, and so on. The data listed in Table 3-13 provides a list of the twelve powerful numbers n with $n \leq 100$.

Expressing the numbers in factored form provides an easier method of determining powerful numbers. For, if you wanted to determine the powerful numbers up to 200, you can list them in the prime factorization form

$$p_1^{\alpha_1} p_2^{\alpha_2} \cdots p_k^{\alpha_k}$$

where the exponents sum first to two, then three, and so on. These numbers are listed in Table 3-14.

Table 3-13. Powerful Numbers ≤ 100.

n	n
$4 = 2^2$	$32 = 2^5$
$8 = 2^3$	$49 = 7^2$
$9 = 3^2$	$64 = 2^6$
$16 = 2^4$	$72 = 2^2 \cdot 3^2$
$25 = 5^2$	$81 = 3^4$
$27 = 3^3$	$100 = 2^2 \cdot 5^2$

Table 3-14. Powerful Numbers ≤ 200.

Sum of Exponents = 2	3	4	5	6	7
2^2	2^3	2^4	2^5	2^6	2^7
3^2	3^3	$2^2 3^2$	$2^2 3^3$	$2^4 3^2$	
5^2	5^3	$2^2 5^2$	$2^3 3^2$		
7^2		$2^2 7^2$			
11^2					
13^2					

You can ask some interesting questions concerning powerful numbers, questions that parallel those on prime numbers. For instance, consider the following.

Question 1: Is there a formula, or expression, that generates powerful numbers?

Question 2: How many powerful numbers are there less than a given value?

Question 3: What proportion of powerful numbers are perfect squares?

Question 4: Are there powerful numbers that are consecutive, and if so, how many?

Question 5: For each positive integer k, can you find two powerful numbers that differ by k? Can you find infinitely many?

Question 4 happens to be satisfied by the two powerful numbers 8 and 9, and also by 288 and 289. But there are more solutions, infinitely many more. For if you set A and B equal to

$$A = 8x_1^2 y_1^2$$
$$B = (x_1^2 + 2y_1^2)^2$$

where x_1 and y_1 are integral solutions to the equation $x^2 - 2y^2 = 1$, then A and B are powerful numbers that differ by one, $B - A = 1$. This follows because

97

$$B - A = (x_1^2 + 2y_1^2)^2 - 8x_1^2y_1^2$$
$$= (1 + 2y_1^2 + 2y_1^2)^2 \ 8(1 + 2y_1^2)y_1^2$$
$$= 1 + 8y_1^2 + 16y_1^4 - 8y_1^2 - 16y_1^4$$
$$= 1.$$

The values $x_1 = y_1 = 1$ produce $A = 8$ and $B = 9$, and $x_1 = 3$, $y_1 = 2$ give $A = 288$ and $B = 289$. The equation $x^2 - 2y^2 = 1$ is a special case of Pell's equation (see any standard text on number theory), which has the standard form $x^2 - Ny^2 = 1$. If N is other than a perfect square, the equation has infinitely many (first stated by Fermat) solutions, and these solutions are obtained from the continued fraction expansion for $\sqrt{N}$.

References to Pell's equation (erroneously named after John Pell, an English mathematician) occur scattered throughout mathematics. The most famous illustration is in the so-called Cattle Problem of Archimedes [57]. The problem appears in poetic form and involves the number of head of cattle of four different colors grazing on the slopes of Sicily. The solution depends on seven equations in eight unknowns, and a couple of side conditions that assert that certain numbers are perfect squares. After some elementary algebra, the problem reduces to solving the equation

$$x^2 - 4,729,494 \ y^2 = 1.$$

The smallest solution is a number y of forty-one digits, although a different interpretation of the problem leads to a solution of some 200,000 digits.

It follows that the powerful numbers are closed under the operation of multiplication. That is, the product of two powerful numbers is again a powerful number. Thus, the product of A and B from above is a powerful number, and so is 4AB. Furthermore, the number that follows this, 4AB + 1, is also powerful since it is a perfect square,

$$4AB + 1 = 4A(A + 1) + 1$$
$$= (2A + 1)^2.$$

Since 288 and 289 are consecutive powerful numbers, so are $4(288)(289) = 332928$ and $339929 = 577^2$.

It was originally thought that if two consecutive numbers are powerful, then the larger number must be a perfect square. This is the case with 8,9 and 288,289 and 9800,9801 and 332928,332929. But this doesn't follow in general because 465124 and 465125 are powerful, and the larger number is not a perfect square (but the smaller one is). And then 12167 and 12168 are powerful, but neither is a perfect square. It is an open question as to how many such examples (without perfect squares) exist.

If you consider question 5, and let k run through the values from 1 to 10, you get the results shown in Table 3-15. It could be that $k = 5$ and $k = 6$ have no representations as the difference of two powerful numbers. You may elect to try and find a representation. How many of the other representations are unique? The infinite number of representations for $k = 1$ have already been discussed. The case $k = 4$ also produces infinitely many representations because if x and $x + 4$ are powerful (such as 4 and 8), then so are $x(x + 4)$ and $x(x + 4) + 4$. Starting the sequence with $x = 4$ gives us the four initial pairs 4 and 8, 32 and 36, 1152 and 1156, and 1331712 and 1331716. It wasn't until recently [68], that a second example was found illustrating two powerful numbers whose difference is two. Apart from $27 - 25 = 2$, this second pair of numbers, n and $n + 2$, occurs when n is quite large; in fact, $n > 70000$. The reference [68] points out that there are infinitely many such representations.

Finally, consider the notion of the density of the powerful numbers. This density number is a measure of what percentage of the whole numbers are powerful. To be more precise, I introduce

Table 3-15. Differences of Powerful Numbers.

k	Powerful Numbers
1	$3^2 - 2^3$
2	$3^3 - 5^2$
3	$2^7 - 5^3$
4	$2^3 - 2^2$
5	
6	
7	$2^4 - 3^2$
8	$2^4 - 2^3$
9	$5^2 - 2^4$
10	$13^3 - 3^7$

tion P(x) to represent the number of powerful numbers that ̱nded above by x. From Table 3-13, you see that P(100) = 12, ̱m Table 3-14 you see that P(200) = 19. To obtain an averaging effect, you need to examine the ratio P(x)/x. Thus

$$\frac{P(100)}{100} = .12 \quad \text{and} \quad \frac{P(200)}{200} = .095.$$

What do you think happens to P(x)/x as x tends to infinity? Does it approach a limiting value? If so, is this limiting value zero, or positive? The density (Schnirelmann density) of the sequence of powerful numbers is defined as the greatest lower bound of the sequence P(n)/n. This numerical value gives you one measure as to how heavily populated the natural numbers are with powerful numbers. Unfortunately the population density is zero, so the powerful numbers are somewhat "scarce." But this is the same density as possessed by the sequence of perfect squares, primes, and powers of two. Amazingly enough, the density of weird numbers (as discussed earlier) is positive!

Since the density of powerful numbers is zero, the limit

$$\lim_{x \to \infty} \frac{P(x)}{x} = 0.$$

Golomb [33] has shown, however, that the ratio $P(x)/\sqrt{x}$ tends to a positive constant c as x tends to infinity,

$$\lim_{x \to \infty} \frac{P(x)}{\sqrt{x}} = c.$$

In effect, this means that for a large x, the number of powerful numbers bounded above by x is approximately equal to $c\sqrt{x}$. Recall that Table 3-13 listed 12 powerful numbers bounded above by 100; thus $12 = c\sqrt{100}$ giving an approximate value of $c = 1.2$. Table 3-14 listed 19 powerful numbers bounded above by 200 thus $19 = c\sqrt{200}$, giving $c = 1.34$.

The approximation to c improves as x increases. Try to improve this approximation by programming the computer to compute and tally the powerful numbers bounded by x, with varying values for x: say $x = 100, 200, 300, \cdots, 1000$. You could check these figures by hand: if they are correct, proceed to have the computer calculate $P(x)$ for $x = 2000, 3000, \cdots, 100000$. The quotients $P(x)/\sqrt{x}$ should be converging to c. Golomb has shown that the exact value of c is related to the zeta function $\zeta(x)$ (to be discussed in Chapter 4) by

$$c = \frac{\zeta(3/2)}{\zeta(3)} \quad .$$

Unfortunately it is a little difficult to evaluate this expression because of the nature of the zeta function.

The project is to compute powerful numbers. What is the best way to do this? Since you are only interested in $x \leq 100000$, you are only interested in those primes $p \leq \sqrt{100000} < 316$. You might need a subprogram to compute these primes, and then it would be best to store them in an array, PRIME(i). You might want to compute the powerful numbers by the method illustrated in Table 3-14 where the numbers are determined by an increasing order on the sum of the exponents on their prime factors. Or you may wish to use the fact that every powerful number can be uniquely represented as the product of a perfect square and a perfect cube,

$$\text{powerful number} = r^2 s^2.$$

Exercises

1. Write a program that determines all the powerful numbers less than or equal to 100000. Print out the values $P(x)$ for $x = 100$, $200, \cdots, 1000$ and $x = 2000, 3000, \cdots, 100000$. Estimate the value of c. Also, find the second pair n, n+2 of consecutive odd powerful numbers.
2. What is the Schnirelmann density of the sequence E_n of even nonnegative integers, $E_n = \{0, 2, 4, 6, \cdots\}$?

FERMAT'S THEOREM

To discuss this section and the next two, I will define what is meant by a congruence, which is surely one of the most fundamental concepts used in number theory.

Definition: If a, b and m are integers, with $m > 0$, then a is said to be congruent to b modulo m, written

$$a \equiv b \pmod{m},$$

if the difference a − b is evenly divisible by m.

It follows that $17 \equiv 3 \pmod 2$ and $118 \equiv -2 \pmod 5$.

An alternative formulation of this definition is that a and b are congruent modulo m (m is the modulus) if a and b both give the same positive remainder when divided by m. Note that 17 and 3 both give a remainder of 1 when divided by 2, and 118 and −2 give a remainder of 3 when divided by 5.

The congruence notation $a \equiv b \pmod{m}$ was invented and first used by the great mathematician Carl Frederick Gauss. The notation is particularly helpful because it enables one to express many properties in short, simple terms. Congruences act like equalities in many ways. The following five properties of congruences tend to bear this out.

Property 1: $a \equiv a \pmod m$

Property 2: If $a \equiv b \pmod m$ then $b \equiv a \pmod m$

Property 3: If $a \equiv b \pmod m$ and $b \equiv c \pmod m$ then $a \equiv c \pmod m$

Property 4: If $a \equiv b \pmod m$ and $c \equiv d \pmod m$ then $a + c \equiv b + d \pmod m$

Property 5: If $a \equiv b \pmod m$ and $c \equiv d \pmod m$ then $ac \equiv bd \pmod m$.

You use the notion of a congruence in everyday life when it comes to reading the time from a clock. Ignoring whether the time is A.M. or P.M., a clock operates on a modulus m = 12 system. For, if it is currently 8:00, then 5 hours later the clock reads 1:00, or

$$8 + 5 \equiv 1 \pmod{12}.$$

The case m = 24 would then differentiate between morning and evening. In some texts this kind of arithmetic is referred to as clock arithmetic.

It is common knowledge that a number N is divisible by 2 if the last digit (units digit) in N is divisible by 2. Furthermore, N is divisible by 4 (and 8) if the number formed by the last two (three) digits is divisible by 4 (and 8). It is also commonly known that N is divisible by 9 if the number formed by summing the digits in N is divisible by 9. N = 123412572 is divisible by 9 because the digits sum to 27, which is a multiple of 9. Why does this method always work? The idea of congruences will show you why. First, since $10 \equiv 1$ (mod 9), it follows from property 5 that $10^2 \equiv 1$ (mod 9), $10^3 \equiv 1$ (mod 9), and, in general, $10^k \equiv 1$ (mod 9). If N is some k + 1 digit number, say

$$N = a_k a_{k-1} \cdots a_3 a_2 a_1 a_0$$

then it follows from property 5 again that

$$
\begin{aligned}
N &= a_k 10^k + a_{k-1} 10^{k-1} + \cdots + a_3 10^3 + a_2 10^2 + a_1 10 + a_0 \\
&\equiv a_k(1) + a_{k-1}(1) + \cdots + a_3(1) + a_2(1) + a_1(1) + a_0 \text{ (mod 9)} \\
&\equiv a_k + a_{k-1} + \cdots + a_3 + a_2 + a_1 + a_0 \text{ (mod 9)}.
\end{aligned}
$$

Now select an arbitrary modulus m and make some computations. These computations will lead you into the celebrated theorem of Fermat. Suppose you first choose m = 12. Then consider values of a ranging from a = 2 to a = 11. For each such choice for a, the powers a^2, a^3, a^4, $\cdots$, a^{12} will be reduced modulo 12. These values are shown in Table 3-16.

A study of this table shows the following. For one thing, if i = 2, you don't have $a^i \equiv 2$ (mod 12) or $a^i \equiv 6$ (mod 12) or $a^i \equiv 10$ (mod 12). But perhaps this might change if a were extended beyond 11; what do you think? On the other hand, for some values of i there are quite a few values of a for which $a^i \equiv a$ (mod 12). In particular, for i = 3, 5, 7, 9, and i = 11 you have $a^i \equiv a$ (mod 12) with the values a = 3, 4, 5, 7, 8, 9, 11. It might be a little difficult to detect a general pattern here. One reason is because m is composite and has a lot of divisors. The factors of 12, or, for that matter, the numbers that share a common factor with 12 tend to camouflage the true relationship. The picture becomes clearer when m is a prime. So set m = 7 and compute a^2, a^3, $\cdots$, a^7 for a = 2, 3, 4, 5, 6 (See Table 3-17).

In this case it is obvious that $a^i \equiv a$ (mod 7) for all a when i = 7. From this you may surmise that $a^p \equiv a$ (mod p) for all a = 2, 3, $\cdots$, (p−1) whenever p is prime. Actually it holds true for any integer a, even if a is a multiple of p. Dividing both sides of the congruence by a gives Fermat's celebrated theorem.

Fermat's Theorem: If p is a prime, and a is any integer, then

$$a^{p-1} \equiv 1 \text{ (mod p)}.$$

Table 3-16. Powers of a Reduced Modulo 12.

a	a^2	a^3	a^4	a^5	a^6
2	4	8	4	8	4
3	9	3	9	3	9
4	4	4	4	4	4
5	1	5	1	5	1
6	0	0	0	0	0
7	1	7	1	7	1
8	4	8	4	8	4
9	9	9	9	9	9
10	4	4	4	4	4
11	1	11	1	11	1

a^7	a^8	a^9	a^{10}	a^{11}	a^{12}
8	4	8	4	8	4
3	9	3	9	3	9
4	4	4	4	4	4
5	1	5	1	5	1
0	0	0	0	0	0
7	1	7	1	7	1
8	4	8	4	8	4
9	9	9	9	9	9
4	4	4	4	4	4
11	1	11	1	11	1

Fermat first stated this theorem in 1640, but did not publish a proof. Perhaps the reason for this is that he was not a professional mathematician; he was instead a French lawyer and judge who indulged in mathematics as a hobby. He probably felt that the discovery gave enough satisfaction and publishing the proof, or even actually proving it, was more than he was inclined to do. It was left to Euler to publish the first proof, some 96 years later in 1736 [36]. Euler generalized the result to cover what happens when the modulus is composite (like m = 12). Here the following theorem fits:

Theorem (Fermat — Euler): Let a, m be positive integers that are relatively prime, (a,m) = 1. If

$\emptyset(m)$ denotes the number of positive integers less than m and relatively prime to m, then

$$a^{\emptyset(m)} \equiv 1 \pmod{m}.$$

The integers 1, 5, 7, and 11 and the four integers less than and relatively prime to $m = 12$, so $\emptyset(12) = 4$. This means $a^4 \equiv 1 \pmod{12}$ for $a = 1, 5, 7, 11, 13, 17, \cdots$.

Let us return to Fermat's Theorem and discuss an interesting programming problem. Suppose you set $p = 17$ and $a = 4$. Verify that $4^{16} \equiv 1 \pmod{17}$. It is not necessary to calculate the exact value of 4^{16} and then divide $4^{16} - 1$ by 17; you can proceed in stages, reducing modulo 17 as you go. You have

$$4^4 = 256 \equiv 1 \pmod{17}$$

so squaring yields

$$4^8 \equiv 1^2 \equiv 1 \pmod{17}$$

and squaring again yields

$$4^{16} \equiv 1^2 \equiv 1 \pmod{17}.$$

Similarly, with $a = 6$ we get

$$6^4 = 1296 \equiv 4 \pmod{17}$$

so

$$6^8 \equiv 4^2 \equiv 16 \equiv -1 \pmod{17}$$

and thus

Table 3-17. Powers of a Reduced Modulo 7.

a	a^2	a^3	a^4	a^5	a^6	a^7
2	4	1	2	4	1	2
3	2	6	4	5	1	3
4	2	1	4	2	1	4
5	4	6	2	3	1	5
6	1	6	1	6	1	6

$$6^{16} \equiv (-1)^2 \equiv 1 \pmod{17}.$$

These two problems were solvable in a few steps because the exponent 16 is a power of 2. But suppose you wished to verify the congruence

$$7^{100} \equiv 1 \pmod{101}.$$

What is the best way to proceed? You need a method that will lend itself to an efficient algorithm. Perhaps the following sequence of steps will outline such an algorithm. With a modulus of 101 you have

$$
\begin{aligned}
7^{100} &\equiv 7^{2 \cdot 50} && \pmod{101} \\
&\equiv (7^2)^{50} && \pmod{101} \\
&\equiv 49^{50} && \pmod{101} \\
&\equiv 49^{2 \cdot 25} && \pmod{101} \\
&\equiv (49^2)^{25} && \pmod{101} \\
&\equiv 78^{25} && \pmod{101} \\
&\equiv 78^{2 \cdot 12 + 1} && \pmod{101} \\
&\equiv (78^2)^{12} 78 && \pmod{101} \\
&\equiv 24^{12} \cdot 78 && \pmod{101} \\
&\equiv (24^2)^6 78 && \pmod{101} \\
&\equiv 71^6 \cdot 78 && \pmod{101} \\
&\equiv (71^2)^3 78 && \pmod{101} \\
&\equiv 92^3 \cdot 78 && \pmod{101} \\
&\equiv 92^2 \cdot 92 \cdot 78 && \pmod{101} \\
&\equiv 92^2 \cdot 5 && \pmod{101} \\
&\equiv 81^1 \cdot 5 && \pmod{101} \\
&\equiv 1 && \pmod{101}.
\end{aligned}
$$

This principle behind this method is that every expression of the form

$$\text{base}^{\text{exponent}} \cdot \text{factor}$$

is reduced modulo p by

$$
\text{base}^{\text{exponent}} \equiv
\begin{cases}
(\text{base}^2)^{\text{exponent}/2} \cdot \text{factor} & \text{if the exponent is even} \\[2mm]
(\text{base}^2)^{(\text{exponent}-1)/2} \text{base} \cdot \text{factor} & \text{if the exponent is odd}
\end{cases}
$$

$$\equiv \quad (\text{new base})^{\text{new exponent}} \cdot (\text{new factor}).$$

106

This process continues until the exponent assumes the value one. Then the product of the base and the factor should be congruent with one modulo p. The flowchart shown in Fig. 3-16 illustrates this procedure of reducing a^{p-1} modulo p.

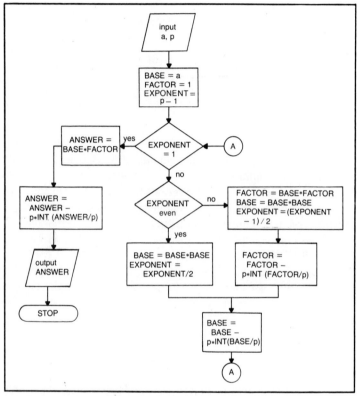

Fig. 3-16. Flowchart verifying $a^{p-1} = 1$ (mod p).

You should be able to fill in the necessary programming details. Of course this isn't the only method for reducing a^{p-1} modulo p and creativity is most certainly encouraged.

Finally, the converse of Fermat's Theorem has evoked much study. This converse is stated: if p divides ($a^p - a$) for every integer a, then p is prime. For a long time this converse was thought to be true; in fact, the particular case with a = 2 was thought to be true (i.e., if p|(2 - 2) then p is prime). People had high hopes that this method would accurately serve to quickly determine whether or not p was prime. Testing p = 1517 for primality you get

$$2^{1516} \equiv 4^{758} \equiv 16^{379} \equiv 305^{99}\ 1062$$
$$\equiv 488^{49}\ 789$$
$$\equiv 625^{12}\ 1231$$
$$\equiv 393^2\ 488$$
$$\equiv 100 \pmod{1517}$$

so 1517 is not prime. But testing p = 341 yields

$$2^{340} \equiv 4^{170} \equiv 16^{85} \equiv 64^{20}$$
$$\equiv 16^5$$
$$\equiv 1 \pmod{341}.$$

To claim, at this point, that p = 341 is prime would be erroneous because 341 factors as 341 = 11•31. This counterexample [46] was not discovered until 1819. The number 341 happens to be the smallest of the infinitely many pseudoprimes, which are those composite numbers n such that n divides $(2^n - 2)$. Pseudoprimes are also commonly known as Poulet numbers [37]; a Poulet number n is said to be a super-Poulet number if every divisor d of n satisfies $d|(2^d - 2)$. The number 2047 is an example of a super-Poulet number because $2047|(2^{2047} - 2)$, and the two divisors 23 and 89 of 2047 both obey $23|(2^{23} - 2)$ and $89|(2^{89} - 2)$. A composite number n that divides $(a^n - a)$ for all a is called an absolute pseudoprime (or Carmichael number). Obviously any absolute pseudoprime (such as 561) is also a pseudoprime, but not vice versa (341 is not an absolute pseudoprime).

Exercises

1. Write a program that verifies $a^{p-1} \equiv 1 \pmod{p}$ for any given value of a and any given prime p. What changes would you make if you wanted to verify the Fermat-Euler Theorem? Test your program with p = 17, a = 2; p = 101, a = 3; p = 101, a = 15.
2. If the number 2^{1314} is multiplied out, what are the last three digits (units, tens, hundreds digits)? Hint: use congruences—do you know what modulus to use?

QUADRATIC CONGRUENCES

Quadratic congruences are congruences of the form

$$ax^2 + bx + c \equiv 0 \pmod{m}$$

where a, b, c, m are integers and x is an integral variable. For the remainder of this section the modulus m will be restricted to being an odd prime. The congruence $2x^2 + 3x + 1 \equiv 0 \pmod 5$ is typical of the quadratic congruences that will be presented.

These congruences are the natural counterparts to the quadratic equations, $ax^2 + bx + c = 0$, so commonly studied in algebra. You shall see that the to "equations" share some important features, yet have some strong differences.

To solve the congruence $ax^2 + bx + c \equiv 0 \pmod m$ means you must seek those integers x that belong to the set S, where

$$S = \{0,\ 1,\ 2,\ \cdots,\ m-1\}$$

and for which $ax^2 + bx + c$ is congruent $0 \pmod m$; implying that m divides $ax^2 + bx + c$. Note that $x = 2$ is a solution to the particular congruence above because 2 is an element of the set $\{0, 1, 2, 3, 4\}$, and $2(2^2) + 3(2) + 1 = 15$ is divisible by 5. The congruence also has a second solution, $x = 4$. There are actually more integers that satisfy the congruence, such as 7, 9, 12, 14, $\cdots$, $5n + 2$, $5n + 4$, but none of these belong to the appropriate set S of allowable solutions.

If it happens that the leading coefficient (the coefficient of x^2) is zero ($a = 0$) and the second coefficient is nonzero ($b \neq 0$), the congruence is linear. Here you have $bx + c \equiv 0 \pmod m$, which is more apt to be written $bx \equiv -c \pmod m$. Since this congruence is equivalent to the equation $bx + c = mn$ for some integer n, you can see by rearranging terms,

$$c = mn - bx,$$

that a necessary and sufficient condition for a solution to exist is that c be divisible by the greatest common divisor of b and m,

$$\gcd(b,m)|c.$$

To illustrate, the congruence $3x \equiv 6 \pmod 3$ has a solution because $\gcd(3,3)|6$, and the congruence $2x \equiv 1 \pmod 5$ has a solution since 2 and 5 are relatively prime ($\gcd(2,5) = 1$), but the congruence $9x \equiv 2 \pmod 3$ is without a solution.

Once you know a solution exists, it is a relatively simple matter to find it. For if $bx \equiv -c \pmod m$ has a solution x, you first check to see if $x = -c/b$ is integral, for if it is, it is the solution. If $-c/b$ is not integral, you then add the modulus m to $-c$ and see if $x = (-c + m)/b$ is integral. In general you continue to try

$$x = (-c + m)/b$$
$$x = (-c + 2m)/b$$
$$\cdot$$
$$\cdot$$
$$\cdot$$
$$x = (-c + km)/b$$

until one of these is integral. Each integral value will be a solution, though they all won't belong to set S. You already know that $2x \equiv 1$ (mod 5) has a solution, but $x = 1/2$ is not it. So you then consider $x = (1 + 5)/2$, which is the integer, $x = 3$, so it is a solution. Next, try $x = (1 + 10)/2$, $x = (1 + 15)/2$ and you will get a second integral value, $x = 8$; but this exceeds the value of the modulus. The congruence $3x \equiv 6 \pmod 3$ happens to have three solutions, $x = 0, 1$, and 2. It follows in general that a linear congruence $bx \equiv -c \pmod m$ will have a solution if and only if $d = \gcd(b,m)$ divides c, and if this is the case, there will be exactly d solutions.

If, on the other hand, the leading coefficient of the quadratic congruence is not equal to zero, which really means $a \not\equiv 0 \pmod m$, an entirely different solution process needs to be formed. The first step in this process is to transform the congruence $ax^2 + bx + c \equiv 0$ (mod m) into an equivalent congruence of the form $x^2 + b'x + c' \equiv 0$ (mod m), where the leading coefficient is one. This is possible because there is an integer a' such that $aa' \equiv 1 \pmod m$. Consequently the congruence $ax^2 + bx + c \equiv 0 \pmod m$ has the same solutions as

$$aa'x^2 + ba'x + ca' \equiv 0 \pmod m$$

which you can now define as

$$x^2 + b'x + c' \equiv 0 \pmod m.$$

Obviously this initial step in the process depends upon being able to determine the integer a'. But this is precisely the same problem as solving a linear congruence, namely $ax \equiv 1 \pmod m$. Since m is prime and $a \not\equiv 0 \pmod m$, it follows that $\gcd(m,a) = 1$, so the linear congruence has a solution; in fact, only one solution. This solution is denoted by a'. To illustrate, the congruence $2x^2 + 3x + 1 \equiv 0$ (mod 5) is equivalent to $x^2 + 4x + 3 \equiv 0$ (mod 5). This follows because the solution to $2a' \equiv 1$ (mod 5) is a' = 3; so

$$
\begin{aligned}
2x^2 + 3x + 1 &\equiv 2 \cdot 3x^2 + 3 \cdot 3x + 1 \cdot 3 \\
&\equiv 6x^2 + 9x + 3 \\
&\equiv 1x^2 + 4x + 3 \ (\text{mod } 5).
\end{aligned}
$$

You should always reduce all the coefficients modulo m, as was done here with the 6 and the 9. This is relatively easy to do; just continue to subtract (or add if the coefficient is negative) m from the coefficient until the difference (sum) is positive but less than m. In general, a number $n > 0$ is reduced modulo m to the following,

$$
n - m * \text{INT (n/m)},
$$

while a negative number, $n < 0$, is reduced to

$$
n + m * \text{INT}(-n/m) + m.
$$

Thus n = 37 is reduced modulo 5 to

$$
37 - 5 * \text{INT}(37/5) \equiv 2
$$

so $37 \equiv 2$ (mod 5), while n = −37 is reduced modulo 5 to

$$
-37 + 5 * \text{INT}(37/5) + 5 \equiv 3
$$

so $-37 \equiv 3$ (mod 5).

The flowchart depicted in Fig. 3-17 furnishes a rough outline of the procedure discussed so far for solving a quadratic congruence.

The second step in the procedure is to complete the square on the congruence. In order to maintain integral coefficients, it is necessary that b' be even. So, if b' is even, then you can complete the square on $x^2 + b'x + c' \equiv 0$ (mod m) and get the following:

$$
\begin{aligned}
x^2 + b'x &\equiv -c' \ (\text{mod } m) \\
x^2 + b'x + (b'/2)^2 &\equiv -c' + (b'/2)^2 \ (\text{mod } m) \\
(x + b'/2)^2 &\equiv -c' + (b'/2)^2 \ (\text{mod } m).
\end{aligned}
$$

$$
\left((x + b'/2)^2 \bmod m \right) = -c' + (b/2)^2
$$

111

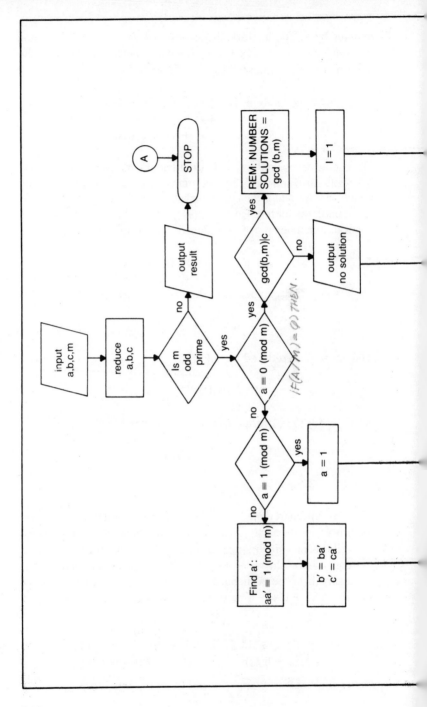

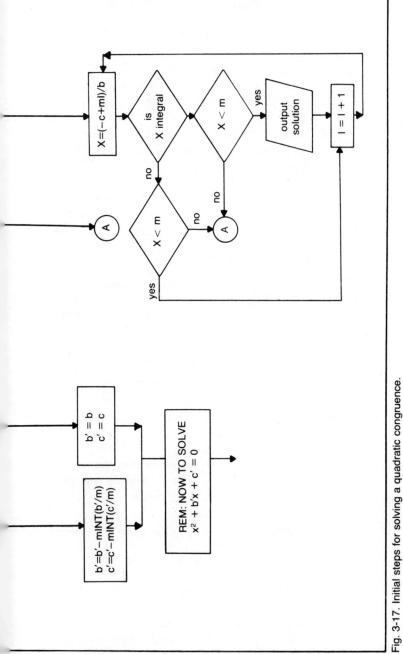

Fig. 3-17. Initial steps for solving a quadratic congruence.

113

This last congruence is a simple quadratic of the form

$$y^2 \equiv e \ (\text{mod } m).$$

On the other hand, if b' is not even, you can change it to $b' + m$, which is even, and then complete the square. Consider the following examples, which illustrate both of these situations.

Example 1: The quadratic $x^2 + 4x + 3 \equiv 0 \ (\text{mod } 5)$ has an even linear coefficient. Completing the square directly gives

$$(x + 4/2)^2 = -3 + (4/2)^2 = -3 + 4 \equiv 1 \ (\text{mod } 5)$$

or

$$y^2 \equiv 1 \ (\text{mod } 5)$$

where $y = x + 2$. There are two solutions to this simple quadratic, namely $y = 1$ and $y = 4$. Hence there are two solutions to the original congruence, these being $(x = y - 2)$

$$x = 1 - 2 = -1 \equiv 4 \ (\text{mod } 5)$$
$$x = 4 - 2 = 2 \equiv 2 \ (\text{mod } 5).$$

As a check, you can see 4 and 2 do satisfy the original congruence.

Example 2: The quadratic $x^2 + 3x + 2 \equiv 0 \ (\text{mod } 5)$ has an odd linear coefficient. So you can rewrite the congruence as $x^2 + 8x + 2 \equiv 0 \ (\text{mod } 5)$, and then complete the square. This yields

$$(x + 8/2)^2 = -2 + (8/2)^2 = 14 \equiv 4 \quad (\text{mod } 5)$$

or

$$y^2 \equiv 4 \ (\text{mod } 5)$$

where $y = x + 4$. This quadratic has two solutions, $y = 4$ and $y = 3$, so the original congruence has the two solutions $x = 2 - 4 = -2 \equiv 3$ and $x = 3 - 4 = -1 \equiv 4 \pmod 5$.

Note here that quadratic congruences can have two solutions just as their counterpart the quadratic equation can. But some equations have no real solutions, and that is the same for quadratic congruences. Consider, for example, the congruence $x^2 + 3x + 3 \equiv 0 \pmod 5$. Completing the square (after increasing the linear coefficient to 8) gives the congruence $y^2 \equiv 3 \pmod 5$. This congruence has no solution! In fact, there are a lot of simple quadratic congruences that lack solutions. Table 3-18 lists some of these congruences.

Table 3-18. Quadratic Congruences that Lack Solutions.

m = 3	m = 5	m = 7	m = 11	m = 13
$y^2 = 2$	$y^2 = 2$	$y^2 = 3$	$y^2 = 2$	$y^2 = 2$
	$y^2 = 3$	$y^2 = 5$	$y^2 = 6$	$y^2 = 5$
		$y^2 = 6$	$y^2 = 7$	$y^2 = 6$
			$y^2 = 8$	$y^2 = 7$
			$y^2 = 10$	$y^2 = 8$
				$y^2 = 11$

For a given odd prime m, it follows that there are $(m - 1)/2$ values of $e \in \{1, 2, \bullet\bullet\bullet, m-1\}$ for which the congruence

$$y^2 \equiv e \pmod m$$

fails to have a solution. Likewise there are this same number $(m - 1)/2$ of values of e for which the congruence does have a solution; and in fact it will have two solutions. If s denotes one such solution, then $m-s$ is the other. You can see, for example, from Table 3-18 that $y^2 \equiv 5 \pmod{11}$ has two solutions; one of them is $y = 4$, so the other must be $11 - 4 = 7$.

The important question is how to determine whether or not the congruence $y^2 \equiv e \pmod m$ has a solution or not. Is this congruence one of the $(m-1)/2$ congruences that does not have a solution? Or is it one of the $(m-1)/2$ congruences that has two solutions? It would be nice to be able to tell the two groups apart. In other words, given the two values of e and m, can you tell whether the congruence $y^2 \equiv e \pmod m$ has a solution? The answer is yes, and this answer is furnished by Euler's Criterion.

Theorem (Euler): If m is an odd prime, and $e \in \{ 1, 2, \cdots,$
(m−1)$\}$, then the congruence $\equiv$
y^2 e (mod m) has a solution if
$$e^{(m-1)/2} \equiv 1 \pmod{m}.$$

The congruence does not have a solution if

$$e^{(m-1)/2} \equiv -1 \pmod{m}.$$

This is a most remarkable result—very simple and elegant. It is even rather amazing that $e^{(m-1)/2}$ is congruent mod m to only 1 and −1 (which is the same as m−1), and not to any other value less than m.

To apply this theorem, you could try to determine whether the congruence $y^2 \equiv 10 \pmod{13}$ has a solution. You must therefore reduce $10^{(13-1)/2}$ modulo 13:

$$10^{(13-1)/2} = 10^6 = 100^3$$
$$\equiv 9^3$$
$$\equiv 81 \cdot 9$$
$$\equiv 3 \cdot 9$$
$$\equiv 1.$$

It follows that the congruence does have a solution; actually two solutions.

You will have to program the reduction of $e^{(m-1)/2}$ and see if it is congruent to +1 or −1. But this is exactly the same program as you needed in the previous section for Fermat's Theorem. Let's run through one final example that incorporates the entire procedure for solving a quadratic congruence.

Example 3: You wish to solve $3x^2 - 7x + 2 \equiv 0 \pmod{37}$. Since $b < 0$, you replace it with 30; this gives $3x^2 + 30x + 2 \equiv 0 \pmod{37}$. The linear equation $3a' \equiv 1 \pmod{37}$ has a solution of $a' = 25$. Multiplying the congruence by 25 gives $75x^2 + 750x + 50 \equiv 0 \pmod{37}$, which reduces to $x^2 + 10x + 13 \equiv 0 \pmod{37}$. Completing the square gives $(x + 5)^2 \equiv 12 \pmod{37}$, and you denote this by $y^2 \equiv 12 \pmod{37}$. This congruence has a solution because

$$12^{(37-1)/2} = 12^{18} = 144^9$$
$$\equiv 33^9$$
$$\equiv 9 \cdot 33$$
$$\equiv 1 \pmod{37}.$$

To find one solution to this, you have to consider which of the numbers 12, 12 + 37, 12 + 37•2, •••, 12 + 37•i is a perfect square. The second one, 12 + 37 = 49 is! So $y = \sqrt{49} = 7$. The second solution must be 37 − 7 = 30. And if x + 5 = y, with y = 7 and 30, x must equal 2 and 25. Thus, the original congruence $3x^2 − 7x + 2 \equiv 0 \pmod{37}$ has the two solutions x = 2, 25.

Exercises

1. Write a program which solves $ax^2 + bx + c \equiv 0 \pmod{m}$. Test your program with the following congruences, keeping in mind that the modulus must be an odd prime:

   a) $3x^2 − 7x + 2 \equiv 0 \pmod{37}$
   b) $3x^2 + 4x + 1 \equiv 0 \pmod{3}$
   c) $2x^2 + 5x + 3 \equiv 0 \pmod{6}$
   d) $2x^2 + 3x + 1 \equiv 0 \pmod{5}$
   e) $x^2 + 0x − 54 \equiv 0 \pmod{7}$.

   It would be a good idea to print out the values of a′, b′, c′, and the simple quadratic $y^2 \equiv e \pmod{m}$ for each problem.

2. Which numbers of the form $n^2 + 1$ are divisible by 7?

QUADRATIC RECIPROCITY

In the previous section on quadratic congruences of the general form $ax^2 + bx + c \equiv 0 \pmod{m}$, it was shown that this congruence can be equivalently expressed as $y^2 \equiv e \pmod{m}$. Since the modulus was understood to be an odd prime, this simple quadratic congruence is more frequently written as

$$x^2 \equiv a \pmod{p}$$

where p is an odd prime, and p / a.

Euler's Criterion states that this congruence is solvable if and only if $a^{(p-1)/2} \equiv 1 \pmod{p}$. If the congruence is solvable, a is called a *quadratic residue* (mod p). On the other hand, if $x^2 \equiv a$ (mod p) has no solution, then a is called a *quadratic nonresidue* (mod p). For example, 3 is a quadratic residue (mod 11), and 5 is a quadratic nonresidue (mod 7). There are also cubic residues, quartic residues, quintic residues, and so on, but the remainder of this section shall only be concerned with quadratic residues. In this case, it is customary to drop the word *quadratic* and just refer to residues and nonresidues. Accordingly, the French mathematician A.M. Legendre introduced the symbol (a/p), known as the *Legendre symbol*, which is defined as being equal to $+1$ or -1, depending on whether a is a residue or nonresidue (mod p);

$$(a/p) = \begin{cases} +1 & \text{if a is a quadratic residue (mod p)} \\ -1 & \text{if a is a quadratic nonresidue (mod p).} \end{cases}$$

From the above examples you can see that $(3/11) = 1$ and $(5/7) = -1$.

It is the purpose of this section to discuss some of the important properties of the Legendre symbol and to show how these properties can be applied to determine whether or not the congruence $x^2 \equiv a$ (mod p) has a solution. Note first that the value of $(1/p)$ is one for all odd prime p. This follows since $x^2 \equiv 1$ (mod p) has the solution $x = 1$. Likewise, the value of $(4/p)$ is one since $x^2 \equiv 4$ (mod p) has the solution $x = 2$. More generally, it follows that $(a^2/p) = 1$ for the same reason.

A second property of the Legendre symbol asserts that if $a \equiv b$ (mod p), then a and b are either both residues or nonresidues (mod p). Equivalently you can write $(a/p) = (b/p)$. This is true because if $(a/p) = 1$, $x^2 \equiv a$ (mod p) has a solution, which is also a solution to the congruence $x^2 \equiv b$ (mod p); therefore, $(b/p) = 1$. But if $(a/p) = -1$, (b/p) must also equal -1 because otherwise $x^2 \equiv b$ (mod p) would have a solution, thus implying that $(a/p) = 1$. Thus, if $a \equiv b$ (mod p), $(a/p) = (b/p)$. This property is useful because it allows you to replace the congruence $x^2 \equiv a$ (mod p) by $x^2 \equiv b$ (mod p) with $a \equiv b$ (mod p). You can see that the two congruences must have identical solution sets. Thus you know that $x^2 \equiv 97$ (mod 13) is equivalent to $x^2 \equiv 6$ (mod 13).

A third property of the Legendre symbol is one that really helps to simplify matters: one that provides added insight towards solving $x^2 \equiv a$ (mod p). Suppose that a is a composite number, say a

$= a_1 a_2$ where a_1 and a_2 are integers other than ± 1. The property states that $(a_1 a_2/p) = (a_1/p)(a_2/p)$. This multiplicative property can be interpreted as saying that the product $(a_1 a_2)$ of two residues (a_1 and a_2) is again a residue; the product of two nonresidues is also a residue, and the product of a residue and a nonresidue is a nonresidue. Thus, because 2 is a nonresidue (mod 13), and 3 is also a nonresidue (mod 13), you can tell that 6 is a residue, so $x^2 \equiv 6$ (mod 13) must have a solution. In other words, $(2/13) = -1$ and $(3/13) = -1$ imply that $(6/13) = +1$. Another viewpoint shows that $x^2 \equiv 6$ (mod 13) has a solution providing $6^{(13-1)/2} \equiv 1$ (mod 13; but

$$6^6 = 2^6 \cdot 3^6 = \begin{cases} +1 \text{ if } 2^6 \equiv 1 \text{ and } 3^6 \equiv 1 \ (\text{mod } 13), \text{ or } 2^6 \equiv -1 \text{ and } 3^6 \equiv -1 \\ \\ -1 \text{ if } 2^6 \equiv 1 \text{ and } 3^6 \equiv -1 \ (\text{mod } 13), \text{ or } 2^6 \equiv -1 \text{ and } 3^6 \equiv 1. \end{cases}$$

So again you can see that $(6/13) = (2/13)(3/13)$.

The multiplicative property can be extended to include composite numbers of k-factors, so that if $a = a_1 a_2 \cdots a_k$ then

$$(a/p) = (a_1/p)(a_2/p) \cdots (a_k/p).$$

Thus, the value of the Legendre symbol $(330/421)$ is equal to the product

$$(2/421)(3/421)(5/421)(11/421).$$

It remains to evaluate each of these four individual symbols, each being of the form (q/p) where q and p are primes. It is the quadratic reciprocity theorem that tells you how (q/p) and (p/q) are related. The theorem was guessed by Euler and Legendre years before it was first proved [11] by Gauss (at the age of 19) in 1796. Gauss was so taken by the theorem that he called it "the gem of the Higher Arithmetic."

Theorem (Quadratic Reciprocity): if p and q are odd primes, and both are congruent to 3 (mod 4), then $(q/p) = -(p/q)$. If, on the other hand, either p or q is congruent to 1 (mod 4), then $(q/p) = (p/q)$.

This theorem allows for tremendous ease in computing Legendre symbols, In the previous problem, for example, you had to compute

(3/421) and (5/421). Because $421 \equiv 1 \pmod 4$, it follows from the reciprocity theorem that

$$(3/421) = (421/3).$$

Furthermore, the second property of the Legendre symbol allows you to say

$$(421/3) = (1/3)$$

because $421 \equiv 1 \pmod 3$. Hence $(1/3) = 1$ from the first property. Similarly you have

$$(5/421) = (421/5) = (1/5) = 1.$$

Isn't this a lot easier than reducing $5^{(421-1)/2}$ modulo 421? In addition you have

$$(11/421) = (421/11) = (3/11)$$

using the same reasoning as above, but now both 3 and 11 are congruent 3 (mod 4), so $(3/11) = -(11/3)$. Applying the second property gives $-(11/3) = -(2/3)$. Consequently the value of $(11/421)$ is the same as $-(2/3)$. Even though this symbol can be easily evaluated by the Euler Criterion

$$(2/3) = -1 \text{ because } 2^{(3-1)/2} \equiv 2^1 \equiv -1 \pmod 3$$

Table 3-19. Values of (2/p).

p	(2/p)	p	(2/p)
3	−1	37	−1
5	−1	41	1
7	1	43	−1
11	−1	47	1
13	−1	53	−1
17	1	59	−1
19	−1	61	−1
23	1	67	−1
29	−1	71	1
31	1	73	1

Table 3-20. Classifying Primes p According to the Value of (2/p).

(2/p) = 1	(2/p) = −1
p	p
7 = 8(0) + 7	3 = 8(0) + 3
17 = 8(2) + 1	5 = 8(0) + 5
23 = 8(2) + 7	11 = 8(1) + 3
31 = 8(3) + 7	13 = 8(1) + 5
41 = 8(5) + 1	19 = 8(2) + 3
47 = 8(5) + 7	29 = 8(3) + 5
71 = 8(8) + 7	37 = 8(4) + 5
73 = 8(9) + 1	43 = 8(5) + 3
	53 = 8(6) + 5
	59 = 8(7) + 3
	61 = 8(7) + 5
	67 = 8(8) + 3

it would be convenient to have an easier rule for evaluating $(2/p)$ when p is any odd prime. Perhaps you can deduce the rule by examining the data listed in Table 3-19, which gives the value of $(2/p)$ for the first 20 odd primes. Those values of p when $(2/p) = 1$ include $p = 7, 17, 23, 31, 41, 47, 71$, and 73. When $(2/p) = -1$, p is any one of $p = 3, 5, 11, 13, 19, 29, 37, 43, 53, 59, 61$ or 67. Rewriting these values of p, as shown in Table 3-20, you can see that $(2/p) = 1$ whenever p is of the form $8k + 1$ or $8k + 7$, while $(2/p) = -1$ when p is of the form $8k + 3$ or $8k + 5$. This gives the following theorem.

Theorem: If p is an odd prime, then

$$(2/p) = 1 \text{ if } p \equiv 1 \pmod{8} \text{ or } p \equiv 7 \pmod{8}$$

$$(2/p) = -1 \text{ is } p \equiv 3 \pmod{8} \text{ or } p \equiv 5 \pmod{8}.$$

Applying this theorem gives $(2/421) = -1$ because $421 \equiv 5 \pmod{8}$.

You can now evaluate any symbol (a/p) with the proper use of these two major theorems and the three properties of the Legendre symbol. Let's work through one final example that requires all five of these results. Furthermore, let us try to follow an algorithmic procedure that lends itself to computer adaption.

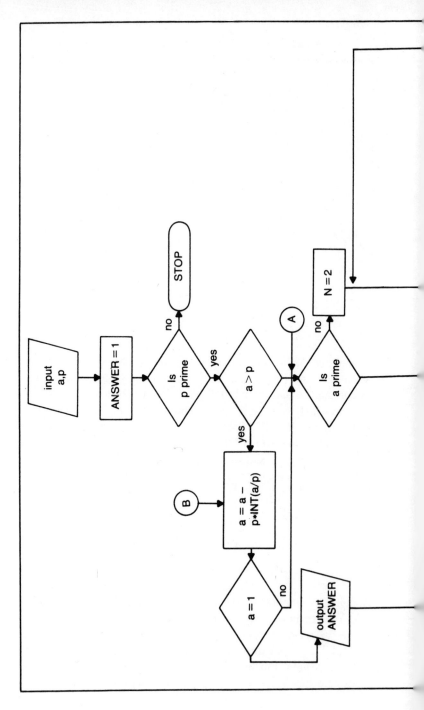

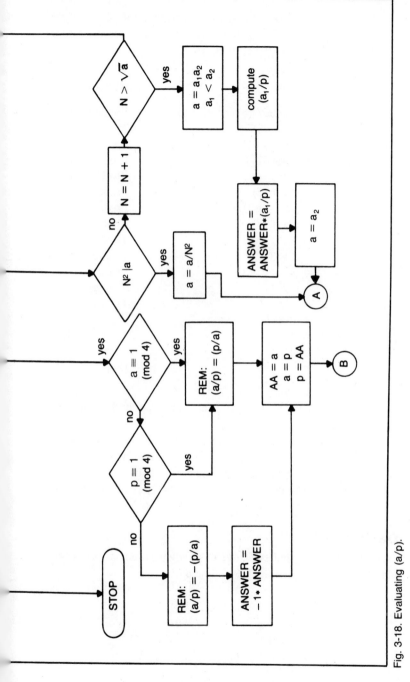

Fig. 3-18. Evaluating (a/p).

Example 1: Evaluate (1234/4567). Check first to be sure that p
= 4567 is prime; it is. Then, you have the following sequence of steps.

Statements		**Reasons**
(1234/4567)	= (2/4567)(617/4567)	Property 3
	= (617/4567)	Theorem
	= (4567/617)	Theorem (QR)
	= (248/617)	Property 2
	= (4/617)(62/617)	Property 3
	= (62/617)	Property 1
	= (2/617)(31/617)	Property 3
	= (31/617)	Theorem
	= (617/31)	Theorem (QR)
	= (28/31)	Property 2
	= (4/31)(7/31)	Property 3
	= (7/31)	Property 1
	= −(31/7)	Theorem (QR)
	= −(3/7)	Property 2
	= (7/3)	Theorem (QR)
	= (1/3)	Property 2
	= 1	Property 1.

Further details on evaluating (a/p) are sketched in the flowchart depicted in Fig. 3-18.

You will need to elaborate on that portion of the flowchart that
requires the computation of (a_1/p) and the determination of the
primality of a and p. Of course, you may elect a different algorithm
for evaluating the Legendre symbol; creativity is greatly encouraged.

Exercises

1. Write a program that evaluates (a/p) when p is an odd prime, and
p∤a. Test your program on the following.

a. (85/97)	c. (1234/4567)
b. (625/9973)	d. (260/71)

2. Show that for each odd prime p, $(2/p) = (-1)^{(p^2-1)/8}$.

124

Chapter 4
Algebra

Algebra signifies for practically all of us that critical point in our education where a variable is first introduced to represent the value of some quantity. The traumatic experience of having to deal with x's and y's, instead of 3's and 4's, is shared by many. It is much harder to comprehend the sum $x^2 + x$ than $7 + 15$. Those that do survive this introduction to abstraction and come to understand it for what it really is are much more fully prepared to cope with problems that crop up in everyday life.

The history of algebra dates back to at least three thousand years ago. Along with geometry, it furnished the Greeks with valuable food for thought. Algebra was, and is, in a sense, the mathematician's language for communicating ideas and results. Students today are more familiar with the famous equation $a^2 + b^2 = c^2$ than they are with the verbal statement of the Pythagorean Theorem, "the square of the hypotenuse is equal to the sum of the squares of the two legs." It is the power of this language that enables people to express ideas more concisely and, at times, more clearly than in any other language.

Like all languages, however, algebra requires much study and thought before you can become proficient in it. First, basic terms are defined, and operational rules are established. Additional results are slowly discovered through a logical process of deduction. Typically a beginner ventures through the simplification of algebraic expressions, to factoring polynomials, to solving linear equa-

and inequalities, to doing quadratic equations and higher order polynomials, and to solving systems of equations. Further topics might include ratio and proportion, exponents and logarithms, permutations and combinations, and determinants.

Advanced courses in algebra cover such diverse topics as groups, commutative rings, integral domains, congruences, homomorphisms and automorphisms, equivalence relations, vector spaces, orthogonal bases and dual spaces, bilinear functions, tensor products, triangular matrices and the characteristic polynomial, Gaussian integers, ideals, and the Galois group. The list goes on and on, but the theme remains constant: the construction of a sound mathematical system containing elements and operations for combining the elements, and then to study the structure of the system. Algebraists are as concerned with the mathematical structure of a number system as chemists are with the molecular arrangement of a compound. It's the algebraists that give meaning to $2 + 4 = 10$ or $2 \cdot 3 = 0$; they show which polynomials can be solved by radicals; they show how determinants can be used to solve systems of equations, and they show that the product $x \cdot y$ is not always equal to $y \cdot x$.

In the beginning, at its lower level, a substantial portion of algebra is devoted to solving equations. Those methods that employ an algorithmic routine solving cubic and quartic equations, are ones that lend themselves quite well to computer assistance. Several of these will be discussed in this chapter.

It is noteworthy that all of the sciences have benefited immensely from proper usage of methods acquired from algebra. The symbolism alone allows for cutting through much red tape. Equations and formulas are integral to all the sciences, and probably each discipline has its favorite formula or expression. For the physicists, it could very well be $E = mc^2$; while for the chemists, it might be $H_2 + 0 = H_2 0$, although their favorite is undoubtedly a much more complicated formula than this. For the mathematicians, though, many stand behind the formula

$$e^{i\pi} + 1 = 0$$

which binds together the five most important constants. A quote attributed to E. Kasner and J. Newman remarks on this formula:

"There is a famous formula—perhaps the most compact and famous of all formulas—developed by Euler from a discovery of deMoivre: $e^{i\pi} + 1 = 0$. It appears equally to the mystic, the scientist, the philosopher and the mathematician. For each it has its own meaning. Though known for over a century, deMoivre's formula came to Benjamin Peirce as something of a revelation. Having discovered it one day, he turned to his students. "Gentlemen," he said, "that is surely true, it is absolutely paradoxical; we cannot understand it, and we don't know what it means, but we have proven it, and therefore we know it must be the truth."

DESCARTES'S RULE OF SIGNS

A substantial portion of a college algebra course is devoted to solving equations, typically polynomial equations such as

$$x^3 + 2x^2 - x - 2 = 0$$

or

$$x^2 - 3x = 4.$$

The usual procedure for solving such equations is to set the expression equal to zero, factor the expression, and then solve each factor. Thus, from above, you would have

$$x^2 - 3x - 4 = 0$$
$$(x - 4)(x + 1) = 0$$

so the solutions to this equation are $4, -1$; and the cubic equation factors to

$$(x - 1)(x + 1)(x + 2) = 0$$

so its solutions are $1, -1, -2$.

Whether or not the student has success solving polynomial equations using this method is clearly dependent upon two factors, namely whether the polynomial factors, and if it does, whether or not the student can correctly ascertain the factors. Whether or not a specific polynomial factors is not an easy question; it is known that the polynomial $p(x)$ has a linear factor $x - a$ if and only if $p(a) = 0$. This idea, known as the factor theorem, is common to most precalculus classes. It serves as a valuable aid in helping to factor the polynomial. For instance, consider the equation,

$$x^3 - 3x^2 + 2x - 6 = 0.$$

If this cubic expression has a linear factor, $x - a$, it must be that $p(a)$ $= 0$ or $a^3 - 3a^2 + 2a - 6 = 0$. Alternately trying $a = \pm 1, \pm 2, \pm 3, \pm 4$ produces the following results,

a	p(a)
1	−6
−1	−12
2	−6
−2	−30
3	0
−3	−66
4	18
−4	−126

Since $p(3) = 0$, you can see that $x - 3$ is a linear factor; thus the cubic factors

$$x^3 - 3x^2 + 2x - 6 = (x - 3) \cdot q(x)$$

where $q(x)$ is some quadratic factor (found by dividing $x - 3$ into the cubic). You would then repeat the procedure with $q(x)$ and try to find a linear factor of $q(x)$ by applying the factor theorem. This factor would also be another factor of the cubic polynomial.

There is some strategy involved in choosing possible values for a. In order to minimize the numbers under consideration, one elementary result from algebra proves handy. This says that if a is a rational solution to the equation $p(x) = 0$, and if you set $a = u/v$, with u and v being integers and the fraction reduced to lowest terms, it must follow that the numerator u divides into the constant term of the polynomial, while the denominator v divides into the leading coefficient. Thus, if a is a rational solution to

$$x^3 - 3x^2 + 2x - 6 = 0$$

then u must divide into -6 (so $u = \pm 1, \pm 2, \pm 3, \pm 6$) and v must divide into 1 ($v = \pm 1$). Consequently, if a is a rational solution, it must be one of the numbers $\pm 1/\pm 1, \pm 2/\pm 1, \pm 3/\pm 1, \pm 6/\pm 1$ or equivalently, $\pm 1, \pm 2, \pm 3, \pm 6$. There was no need to test $a = \pm 4$; that was essentially a waste of time.

Likewise, the only possible rational solutions to the equation

$$2x^4 + 5x^3 + 3x^2 + 15x - 9 = 0$$

are those where a = u/v with u = ±1, ±3, ±9 and v = ±1, ±2. In other words, the possible values for a include ±1, ±3, ±9, ±1/2, ±3/2, and ±9/2.

Of course it is convenient to have rational solutions to polynomial equations. Students have a better comprehension of a rational (especially integral) root than an irrational root, and fortunately, most of the equations in texts have rational solutions. Again, though, if you conjecture that an irrational number a is a solution to the equation p(x) = 0, proof of this will depend simply on whether p(a) = 0.

In summary, two techniques from algebra that are very useful in solving polynomial equations have been discussed. This section will now present a third result, one that will help you narrow down the set of possible values for solutions. This result is known as Descartes's rule of signs.

If you consider the polynomial equation p(x) = 0, Descartes's rule gives you the following two results concerning how many positive and negative solutions there are. It says nothing about the explicit value of the solutions.

Result 1: The difference between the number of variations V^+ of the sign of p(x) and the number of positive solutions is an even nonnegative integer.

Result 2: The difference between the number of variations V^- of the sign of p(−x) and the number of negative solutions to p(x) = 0 is an even nonnegative integer.

These results can be illustrated by considering the equation

$$2x^4 + 5x^3 + 3x^2 + 15x - 9 = 0.$$

Since the signs of p(x) are + + + + −, the value of V^+ is one. Hence the number of positive solutions must be one. Similarly, the signs of $p(-x) = 2(-x)^4 + 5(-x)^3 + 3(-x)^2 + 15(-x) - 9 = 2x^4 - 5x^3 + 3x^2 - 15x - 9$ are + − + − −, so $V^- = 3$, which means the number of negative solutions to the original equation is either 1 or 3. Knowing the number of positive and negative solutions makes your work a lot easier, especially when you are testing the rational possibilities.

Any coefficients of p(x) that are zero are automatically excluded from consideration of sign changes. You only count the variations from + (plus) to − (minus), or minus to plus. The examples in Table 4-1 help to clarify this.

Table 4-1. Variations of the Sign.

Polynomials	V⁺	V⁻
$x^3 - 3x^2 + 2x - 6$	3	0
$x^3 + 2x^2 - x - 2$	1	2
$x^3 + 2x^2 - 2$	1	2
$x^6 - x^4 - 4x^2 + 4$	2	2

Consider the last polynomial, $p(x) = x^6 - x^4 - 4x^2 + 4$. If you were to feed this polynomial into the computer, you would have to input all seven coefficients, $1,0,-1,0,-4,0,4$, and it would be helpful to input $n = 6$, the degree of the polynomial. The coefficients could be stored in array A, with

$$A(0) = 4, A(1) = 0, A(2) = -4,$$
$$A(3) = 0, A(4) = -1, A(5) = 0, A(6)$$

You will need to program computation of V⁺ (VPLUS). Examine the nonzero coefficients, and check their signs for a variation change. There are several ways this can be done: one method is outlined in the flowchart depicted in Fig. 4-1.

The technique used is to place the nonzero elements from A in array B, and then to check whether or not consecutive values $B(i)$ and $B(i + 1)$ are of opposite signs: $B(i)*B(i+1) < 0$. For example, using the values for A from above, you would have $B(1) = 4$, $B(2) = -4$, $B(3) = -1$ and $B(4) = 1$. The final value of i would be 4 (that tells us how many terms are in B) and the final value of VPLUS would be 2.

Now you need to compute V⁻ (VMINUS). That means you need a new set of coefficients, or a new set of values for A. But the only changes will be for those coefficients of x raised to an exponent that is odd. This means that the new coefficient $A(i)$ will equal the negative of the old coefficient $A(i)$ when i is odd; otherwise the values are the same. Consequently, you could add the following set of statements to the program,

```
100   FOR I = 1, N, 2
110      A(I) = -1*A(I)
120   NEXT I
```

and then the computation of VMINUS follows the same procedure as that of VPLUS. You might want to maintain the original coefficients, so you could simply define a new array, say AA, to store the coefficients of p(−x). This would be done as follows.

```
100   FOR I = 0, N, 2
110      AA(I) = A(I)
120      AA(I + 1) = -1*A(I)
130   NEXT I
```

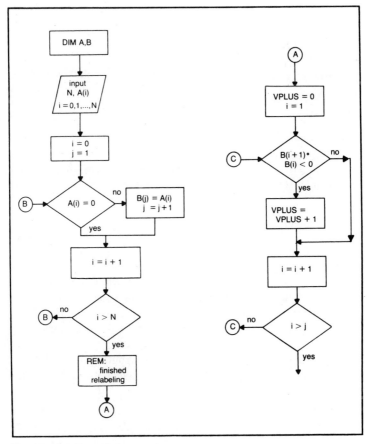

Fig. 4-1. Variations of sign.

Then, again, computing VMINUS would follow the same path as VPLUS.

Recent investigation [22] into Descartes's rule has shed some insight on to how many complex solutions $p(x) = 0$ has. This number is a function of how many zero coefficients the polynomial p has and how they are grouped together with the nonzero coefficients.

To study this, set $p(x) = a_n x^n + a_{n-1} x^{n-1} + \cdots + a_1 x + a_0$, and define a gap in p to be a pair (a_i, a_j) of nonzero coefficients where all the coefficients in between a_i and a_j are zero. The polynomial $p(x) = x^6 + 4x^3 - 7x + 1$ has two gaps, denoted by $(1,4)$ and $(4,-7)$. The gap (a_i, a_j) is said to be even when there is an even number of zero coefficients between a_i and a_j, and odd when there is an odd number. Thus, $(1,4)$ is an even gap, while $(4,-7)$ is an odd gap. Finally, the gap (a_i, a_j) is called a performance when a_i and a_j have the same sign (i.e., $a_i a_j > 0$), and the gap is called a variation when a_i and a_j have opposite signs ($a_i a_j < 0$). The gap $(1,4)$ is a performance, while $(4,-7)$ is a variation.

Now measure each gap according to some rule, sum all the measures, and call this number, M, the measure of the polynomial. The equation $p(x) = 0$ will then have at least M complex solutions!

The measure of any even gap is that even number of zero coefficients between a_i and a_j. The measure of any odd permanence gap is one more than the odd number of zero coefficients between a_i and a_j, while the measure of any odd variation gap is one less than the odd number. The values shown in Table 4-2 will help to illustrate these points.

If you consider the equation $x^9 - 3x^2 + 1 = 0$, you know that, since $M = 6$, the equation has at least 6 complex solutions and at most 3 real solutions. But $V^+ = 2$ and $V^- = 1$, so there is precisely one negative root, and either 0 or 2 positive roots. The polynomial $p(x) = x^9 - 3x^2 + 1$ crosses the x-axis somewhere between 0 and 1 ($p(0)) = 1$. $p(1) = -1$) by the Intermediate Value Theorem; thus there must be exactly 2 positive solutions. None of the 3 real solutions are rational, because the only candidates for rational roots are ± 1.

Extend the program outlined for computing V^+ and V^- to include the evaluation of M. This is the essence of your task in the following problem.

Exercises

1. Write a program that will evaluate V^+, V^- and M for a given polynomial p. You may wish to print out all the gaps together

Table 4-2. The Measures of Particular Gaps and Polynomials.

M	p(x)	Gaps	Type	Gap Measure
2	$x^6 + 4x^3 - 7x + 1$	$(1,4)$	even, permanence	2
		$(4,-7)$	odd, variation	0
4	$7x^8 + 2x^6 - x^4 + x - 1$	$(7,2)$	odd, permanence	2
		$(2,-1)$	odd, variation	0
		$(-1,1)$	even, permanence	2
6	$x^9 - 3x^2 + 1$	$(1,-3)$	even, variation	6
		$(-3,1)$	odd, variation	0

Table 4-3. Illustrating Luddhar's Second Result.

p(x)	l	m	p	q	Factors
$6x^3 - 11x^2 - 4$	-2	2	1	-12	$(-1/2)(x-2)(-12x^2 - 2x - 4)$
$4x^3 - x - 12$	-9	8	6	-6	$(-1/54)(6x-9)(-36x^2 - 54x - 72)$

133

with their individual measures. Test your program with the sample polynomial $p(x) = 7x^9 - 4x^7 + 3x^3 + 2x - 1$.

2. Does the equation

$$\frac{x^4 + 2x + 1}{x - 3} = x + 1$$

have any complex roots?

LUDDHAR'S METHOD FOR SOLVING CUBICS

This section deals with the cubic equation of the form

$$ax^3 + bx^2 + cx + d = 0$$

where the coefficients a, b, c, d are all integers. The integers could be positive, negative or zero, but the leading coefficient must be nonzero, $a \neq 0$. Thus you may wish to examine $2x^3 - 7x^2 + 4x - 5 = 0$ or $x^3 + 2x - 1 = 0$, but you will not consider $x^3 - 4x^2 + x - \sqrt{2} = 0$ nor $2x^3 + \pi x^2 + 1 = 0$. If you are interested in solving cubics of this latter type, where the coefficients are free to assume any real value, you may wish to consult [21] where the general cubic formula (the counterpart to the quadratic formula) is discussed.

When solving any polynomial equation, it is desirable that the polynomial factors with integral coefficients. In our case, with the polynomial $p(x) = ax^3 + bx^2 + cx + d$, any one of four things can happen in relationship to how the polynomial factors; they are

1. the polynomial may not factor
2. one of the factors is a monomial
3. one factor is a linear factor $(ex + f)$ with two nonzero coefficients and the other is a quadratic factor $(gx^2 + hx + i)$ with three nonzero coefficients
4. one factor is a linear factor with two nonzero coefficients, and the other is a quadratic with a missing x-term $(gx^2 + i)$.

The remainder of this section will concentrate solely on items 3 and 4. The results that follow have been credited to Hari Ram Luddhar [50].

First the fourth possibility will be examined. Since $p(x)$ factors as

$$ax^3 + bx^2 + cx + d = (ex + f)(gx^2 + i)$$

134

by multiplying the right-hand side you get

$$ax^3 + bx^2 + cx + d = (eg)x^3 + (fg)x^2 + (ei)x + (fi)$$

Because these are identical polynomials, the corresponding coefficients must be equal

$$a = eg, \qquad c = ei$$
$$b = fg, \qquad d = fi.$$

It follows that the ratio of a to b is equal to the ratio of c to d,

$$\frac{a}{b} = \frac{eg}{fg} = \frac{e}{f} = \frac{ei}{fi} = \frac{c}{d}\,.$$

Thus, the first result of Luddhar can be easily stated.

Result 1:   The cubic polynomial $p(x) = ax^3 + bx^2 + cx + d$ can be expressed as a product of a linear factor and a quadratic factor, with a missing x-term, if and only if the ratio a/b equals c/d. In this case, the equation $p(x) = 0$ has a rational root located at $-b/a$.

This result can be illustrated with the sample $4x^3 - x^2 + 8x - 2 = 0$. Since $4/(-1) = 8/(-2)$, you know that the cubic factors as the product of a linear and a quadratic

$$4x^3 - x^2 + 8x - 2 = x^2(4x - 1) + 2(4x - 1)$$
$$= (4x - 1)(x^2 + 2).$$

From this you can see that the only real root to the equation is 1/4, and the two imaginary roots are $\pm \sqrt{2}\, i$.

Since you know that under these conditions the cubic equation always has a rational root at $-b/a$, one question that pops up is, "Could there be more rational, or real, roots? The other two roots don't have to be complex, do they?" The answer is no, they don't have to be complex, as seen by

$$0 = 4x^3 - x^2 - 8x + 2 = (4x - 1)(x^2 - 2)$$

which has three real roots. Can you construct a cubic equation of this type with three rational roots?

Furthermore, it would be convenient if the computer would print out the linear (ex + f) and quadratic ($gx^2 + i$) factors. How can you program it to do so? The values of a, b, c, d are known because they are input data, but you don't know the values of e, f, g, or i. The following relationships, however, are known:

1. a = eg   2. b = fg   3. c = ei   4. d = fi
5. a/b = e/f = c/d.

These relationships imply that if you know the values of g and i you can compute the values of e and f. This follows since e = a/g = c/i and f = b/g = d/i. Actually though, if you only knew g, you could compute all e, f, and i using the relationships,

6. e = a/g
7. f = b/g
8. i = d/f = dg/b.

Since the values of all these unknowns, in particular g, are integral, you can run g through a loop of possible values ± 1, ± 2, ± 3, •••, ± N and test whether or not the values of e, f, and i as determined in 6, 7, and 8 satisfy relationships 1 through 5. This would give you the factors of $ax^3 + bx^2 + cx + d$, when a/b = c/d.

Luddhar's second result concerns factoring $p(x) = ax^3 + bx^2 + cx + d$ into a linear and a full quadratic. Assume that a ≠ 0 (otherwise you would not have a cubic) and d ≠ 0 (otherwise you would have a monomial factor). This second result is stated slightly differently than the first result was.

Result 2. The equation $ax^3 + bx^2 + cx + d = 0$ with a ≠ 0 and d ≠ 0, has a rational root if there exists the nonzero integers l, m, p, q such that a = pq/l, b = p + q, c = l + m, and d = lm/p. The rational root is −l/p, and the cubic factors as (px + l) ($pqx^2 + plx + lm$)/pl.

Because the relationships involve all eight parameters, the equation $ax^3 + bx^2 + cx + d = 0$ can be written

$$\frac{0}{l} = pq \; x^3 + (p + q)x^2 + (l + m)x + \frac{lm}{p}$$

136

and multiplying both sides by pl gives

$$0 = p^2qx^3 + (p^2l + pql)x^2 + (pl^2 + plm)x + l^2m$$

which factors to

$$0 = (px + l)(pqx^2 + lpx + lm).$$

This means that this equation and the original equation are equivalent; they have the same roots, of which one is clearly $-l/p$. The other roots could be found by solving the quadratic, but you are not really concerned with them at this point. You are more interested in determining the two factors and the one rational root.

Now let's see how you might apply this result. Suppose you wish to solve $2x^3 + 5x^2 + x - 3 = 0$. Typically the first step in the procedure is to select the two nonzero integers l and m whose sum is c. Thus, since $c = 1$, you seek l and m such that $l + m = 1$. Obviously there are many choices for l and m, and if you were writing a program to simulate these actions, it might be best to run l through a loop, allowing it to assume values from 1 to some upper bound, say BOUND, and from $-$BOUND to -1. The value chosen for BOUND probably won't need to be any greater than 50. It follows that the variable m is determined by $m = c - l$. So, you might select to consider the values in the following order,

$$l = 1, \qquad m = c - 1$$
$$l = -1, \qquad m = c + 1$$
$$l = 2, \qquad m = c - 2$$
$$l = -2, \qquad m = c + 2$$
$$\bullet$$
$$\bullet$$
$$\bullet$$
$$l = \text{BOUND}, \qquad m = c - \text{BOUND}$$
$$l = -\text{BOUND}, \qquad m = c + \text{BOUND}$$

because of the monotonicity of absolute l.

This initial part of the program can be outlined by the following flowchart shown in Fig. 4-2. For this particular example with $2x^3 + 5x^2\ x - 3 = 0$, we set $l = 3$ and $m = -2$ (because these are the values that work).

The second step in the procedure involves the relationship $d = lm/p$. Having selected a pair of integers for l and m, the value of p is

137

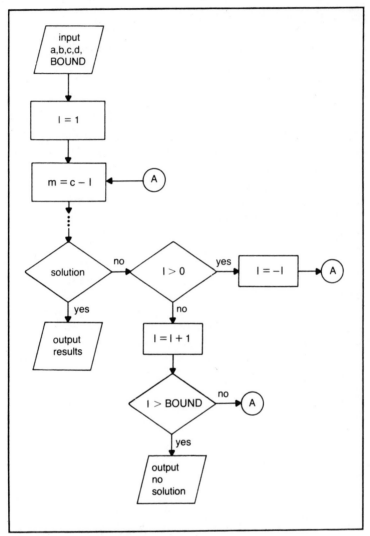

Fig. 4-2. Initial steps in Luddhar's method.

now determined by $p = lm/d$. Keep in mind that p must be an integer. You could therefore refine the flowchart in Fig. 4-2 by the insertions shown in Fig. 4-3. In this particular example, since $d = -3$, it follows that $p = -6/-3 = 2$.

The third step in the procedure is to set q equal to the difference $b - p$ and then check the condition $a = pq/l$. If everything

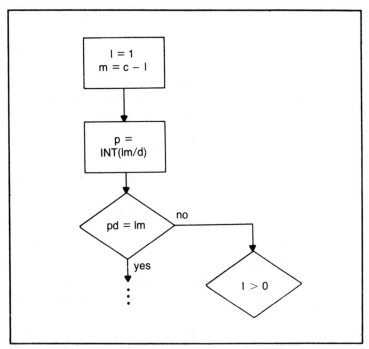

Fig. 4-3. Luddhar's method refined.

works up to this stage, you have obtained a rational root of the equation. Finally, in regards to this example, with $p = 2$, it follows that $q = b - p = 5 - 2 = 3$. When you check $2 = 2(3)/3$, you find you are finished. You should be able to finish the flowchart by completing this last step.

Luddhar's result assures us that $-l/p$ is a rational root and that the cubic $ax^3 + bx^2 + cx + d$ factors as

$$ax^3 + bx^2 + cx + d = \frac{(px + l)(pqx^2 + lpx + lm)}{pl}.$$

Thus, you know

$$2x^3 + 5x^2 + x - 3 = \frac{(2x + 3)(6x^2 + 6x - 6)}{6}$$

and so $-3/2$ is a rational root.

Two other examples that illustrate this second result of Luddhar's method are summarized in Table 4-3.

Exercises

1. Write a program that allows you to input $p(x) = ax^3 + bx^2 + cx + d$ and find out if you can apply either of Luddhar's two results to the equation $p(x) = 0$. If the cubic satisfies the appropriate conditions, program the computer to print out the linear and quadratic factors, the values of the new coefficients, and the rational root. Test your program with the following cubic equations.

$$1. \ 6x^3 - 2x^2 + 9x - 3 = 0$$
$$2. \ 48x^3 - 40x^2 - 21x + 18 = 0$$
$$3. \ x^3 + 2x^2 + 3x + 4 = 0.$$

2. Since $x^3 + 3x^2 + 3x + 1 = (x + 1)(x^2 + 2x + 1)$, what are the values of l, m, p, q?

QUARTIC EQUATIONS

The purpose in this section is to discuss a procedure that will allow you to find all the real roots to the quartic equation

$$x^4 + a_3x^3 + a_2x^2 + a_1x + a_0 = 0$$

where the coefficients are real numbers. Note that any quartic equation can be put in this form by dividing through by the leading coefficient.

Descartes's method [17], which applies to those quartic equations with $a_3 = 0$, will be discussed first. The equation above can be transformed into what is called the *reduced quartic equation* by the substitution $z = x + a_3/4$: thus

$$(z - a_3/4)^4 + a_3(z - a_3/4)^3 + a_2 (z - a_3/4)^2 + a_1(z - a_3/4) + a_0 = 0$$

reduces to an equation of the form

$$z^4 + b_2z^2 + b_1z + b_0 = 0$$

where the coefficients satisfy

$$b_2 = a_2 - (6/16)a_3^2$$
$$b_1 = a_1 - (a_2a_3)/2 + (1/8)a_3^3$$
$$b_0 = a_0 - (a_1a_3)/4 + (a_2a_3^2)/16 - (3/256)a_3^4.$$

Note that this quartic is reduced because of the absence of the z^3 term. As it stands now, you would input the values a_3, a_2, a_1, and a_0 into the computer, along with the defining relations for b_2, b_1, and b_0, and the computer could output the reduced quartic equation.

Descartes's method continues by noting that the reduced quartic factors as a product of two quadratics

$$z^4 + b_2 z^2 + b_1 z + b_0 = (z^2 + kz + h_1)(z^2 - kz + h_2)$$

where, because of the absence of a z^3 term, the two linear coefficients are of the same absolute value, yet opposite in sign. These coefficients are denoted by k and $-k$. Furthermore, if $b_1 = 0$, the quartic equation is actually a quadratic equation,

$$(z^2)^2 + b_2(z^2) + b_0 = 0$$

in z^2, and the quadratic formula can be used to solve for z^2 by taking square roots then giving the value of z. If $b_1 \neq 0$ you must continue using Descartes's method, as shown in Fig. 4-4.

Multiplying the two quadratics gives

$$z^4 + (h_1 + h_2 - k^2)z^2 + k(h_2 - h_1)z + h_2 h_1$$

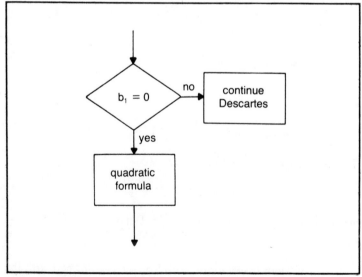

Fig. 4-4. A decision statement in Descartes's method.

and equating coefficients yields

$$b_2 = h_1 + h_2 - k^2$$
$$b_1 = k(h_2 - h_1)$$
$$b_0 = h_2 h_1.$$

Since we have assumed $b_1 \neq 0$ this implies (from the middle equation) that $k \neq 0$, which is good because shortly you will be dividing by k. Multiply both sides of the first equation by k and then the first two equations can be rewritten as

$$h_1 k + h_2 k = k^3 + b_2 k$$
$$-h_1 k + h_2 k = b_1.$$

Solving this system for h_1 and h_2 gives

$$h_1 = \frac{k^3 + b_2 k - b_1}{2k}$$

$$h_2 = \frac{k^3 + b_2 k + b_1}{2k}.$$

Therefore, using the third equation,

$$b_0 = h_2 h_1 = \frac{(k^3 + b_2 k + b_1)}{2k} \frac{(k^3 + b_2 k - b_1)}{2k}$$

or

$$4b_0 k^2 = (k^3 + b_2 k + b_1)(k^3 + b_2 k - b_1).$$

This equality can be written as

$$k^6 + 2b_2 k^4 + (b_2{}^2 - 4b_0)k^2 - b_1{}^2 = 0$$

which is actually a cubic equation in k^2! Denote this cubic by $w^3 + c_2 w^2 + c_1 w + c_0 = 0$ where $w = k^2, c_2 = 2b_2, c_1 = b_2{}^2 - 4b_0$, and $c_0 = -b_1{}^2$.

It may seem that a lot of work has been performed with the result that you've only gotten yourself into a deeper mess, because of a more complicated equation, than when you started. But this isn't the case; progress is being made. The original problem was to solve the quartic

$$x^4 + a_3x^3 + a_2x^2 + a_1x + a_0 = 0.$$

This quartic will be solved as soon as you solve the cubic

$$w^3 + c_2w^2 + c_1w + c_0 = 0$$

because, once you have a root for w (all you need is one value for w, not all three), then you can compute k. Knowing k allows you to determine h_1 and h_2, which gives you the two quadratic factors of the reduced quartic. Thus, you can determine all the real z roots, which then gives you all the x roots. (Recall that $x = z - a_3/4$.) So, all you need to do is solve the general cubic equation

$$w^3 + c_2w^2 + c_1w + c_0 = 0.$$

It won't suffice here to just search for rational roots, as was discussed in the previous sections, because most cubics don't have any rational roots. This is especially applicable here because the coefficients c_0, c_1, and c_2 are stored in the computer as decimals (possibly rounded off) rather than as integers.

To begin, make the substitution $y = w + c_2/3$, which gives

$$(y - c_2/3)^3 + c_2(y - c_2/3)^2 + c_1(y - c_2/3) + c_0 = 0$$

or, what is known as the *reduced cubic,*

$$y^3 + d_1y + d_0 = 0$$

where

$$d_1 = c_1 - c_2^2/3$$
$$d_0 = c_0 - c_1c_2/3 + 2c_2^3/27.$$

The three solutions to the reduced cubic are either real or complex (either one real, or three reals) and can be exactly expressed by

1. $P + Q$

2. $-\left[\dfrac{P + Q}{2}\right] + \left[\dfrac{P - Q}{2}\right]\sqrt{-3}$

3. $-\left[\dfrac{P + Q}{2}\right] - \left[\dfrac{P - Q}{2}\right]\sqrt{-3}$

where the numbers P and Q are defined by

$$P = \sqrt[3]{-d_0/2 + \sqrt{d_0^2/4 + d_1^3/27}}$$

$$Q = \sqrt[3]{-d_0/2 - \sqrt{d_0^2/4 + d_1^3/27}}.$$

There is no problem in taking the cube root of a number, but you must be careful that the radicand of the innermost radical is not negative. It actually follows that

    a. If $d_0^2/4 + d_1^3/27 > 0$ the reduced cubic equation will have exactly one real root, namely $P + Q$, and two imaginary roots.

    b. If $d_0^2/4 + d_1^3/27 = 0$ the reduced cubic equation will have three real roots of which at least two are equal. The real roots are $P + Q$ and $-(P + Q)/2$.

    c. If $d_0^2/4 + d_1^3/27 < 0$ the reduced cubic will have three real and unequal roots.

The third case requires a trigonometric solution in order for you to get a handle on the three roots. Define the angle θ by the formula

$$\theta = \cos^{-1}(-d_0/2 \div \sqrt{-d_1^3/27}).$$

The computer will most likely have the subprogram for ARCCOS built in, or if it doesn't, it will have ARCTAN (which might be denoted by ATN). In this case, if you set ARGU $= -d_0/2 \div \sqrt{-d_1^3/27}$, the angle θ is found by

$$\theta = \text{ARCTAN} (\sqrt{1 - \text{ARGU*ARGU}} \, / \, \text{ARGU}).$$

The value for θ will be given in radian measures. Finally, the three real solutions to the reduced cubic from (c) are

1. $2\cos(\theta/3)\sqrt{-d_1/3}$

2. $2\cos(2\pi/3 + \theta/3)\sqrt{-d_1/3}$

3. $2\cos(4\pi/3 + \theta/3)\sqrt{-d_1/3}.$

Once again, knowing the solution to the reduced cubic allows you to backtrack all the way back and solve the original quartic equation. Let's run through a couple of examples and follow the entire procedure.

Example 1. Consider the equation $f(x) = 0$ where $f(x) = x^4 + 4x^3 + x - 1$. The substitution $z = x + 1$ produces the reduced quartic

$$z^4 - 6z^2 + 9z - 5 = 0.$$

This quartic factors as

$$z^4 - 6z^2 + 9z - 5 = (z^2 + kz + h_1)(z^2 - kz + h_2)$$

where $k^6 - 12k^4 + 56k^2 - 81 = 0$.
You, therefore, need to solve the cubic $w^3 - 12w^2 + 56w - 81 = 0$. The substitution $y = w - 4$ produces the reduced cubic

$$y^3 + 8y + 15 = 0.$$

The expression $d_0^2/4 + d_1^3/27$ is approximately 75.21296, so you know the reduced cubic as only one real root, namely

$$y = P + Q = 1.05449 - 2.52887 = -1.47438.$$

This means $w = 2.52562$, and since $k^2 = w$ you know (the positive square root is sufficient) that $k = 1.58922$. Because $h_1 = -4.56877$ and $h_2 = 1.09439$, the reduced quartic factors as

$$(z^2 + 1.58922z - 4.56877)$$
$$(z^2 - 1.58922z + 1.09439).$$

These roots are $z = -3.075$, $z = 1.48578$, and $z = .7946 \pm .6804i$, so the only real x roots are $x = -4.075$ and $x = .48578$. The results are quite accurate because checking gives $f(-4.075) = .000094$ and $f(.48578) = .0000094$.

Example 2. Consider $g(x) = 0$ where $g(x) = x^4 - 5x^3 + 5x^3 + 5x - 6$, so $a_0 = -6$, $a_1 = 5$, $a_2 = 5$ and $a_3 = -5$. The substitution $z = x - 5/4$

gives the reduced quartic with $b_0 = 189/256$, $b_1 = 15/8$ and $b_2 = -35/8$. The value of k must satisfy

$$k^6 - \frac{35}{4} k^4 + \frac{1036}{64} k^2 - \frac{225}{64} = 0.$$

Thus, to solve $w^3 - 35w^2/4 + 1036w/64 - 225/64 = 0$, you must make the substitution $y = w - 35/12$ and get the reduced cubic with $d_1 = -9.33333$ and $d_0 = -5.92593$. But $d_0^2/4 + d_1^3/27 < 0$, so there are three real roots to the reduced cubic. Since $\theta = \cos^{-1}(.5399496) = 1.000419$ the three roots are,

$$\text{root 1} = 3.3333333$$
$$\text{root 2} = 3.5276684$$
$$\text{root 3} = -.66668.$$

Selecting one of them, say $y = \text{root 1}$, you can compute w by $w = y + 35/12$. (You must make sure $w > 0$, otherwise you should select a different value for y.) The result will be $w = 6.25$. This gives $k = 2.5$, $h_1 = .5625$, and $h_2 = 1.3125$. You may wish to perform a quick check on the work done so far, by showing that $h_1 h_2 = b_0$. The reduced quartic must factor as

$$(z^2 + 2.5z + .5625)(z^2 - 2.5z + 1.3125)$$

from which you get $z = -.25, -2.25, .75,$ and 1.75. This means you have four real x roots, namely $x = 1$, $-1, 2, 3$. Nice results!!

In summary, the solution to a general quartic can be reduced to follow an algorithm of 11 major steps. These are

1. input a_0, a_1, a_2, a_3
2. compute b_0, b_1, b_2
3. compute c_0, c_1, c_2
4. compute d_0, d_1
5. solve reduced cubic for y (only one needed)
6. solve for w (make sure $w > 0$)
7. solve for k (only one needed)
8. solve for h_1, h_2
9. check: $h_1 h_2 = b_0$
10. solve reduced quartic for z (find all real roots)
11. solve for x

146

Exercises

1. Write a program that uses the method on the preceding page to compute all real solutions to $x^4 + a_3x^3 + a_2x^2 + a_1x + a_0 = 0$. Output all key data. Check your program with the following equations.

   1. $x^4 - 4x^3 + 6x^2 - 4x + 1 = 0$ real roots: 1
   2. $x^4 - 10x^3 + 35x^2 - 50x + 24 = 0$ real roots: 1,2,3,4
   3. $x^4 - 2x^2 + 1 = 0$ real roots: 1,1,−1,−1
   4. $x^4 + x^3 - 3x^2 - x + 2 = 0$ real roots: 1,1,−1,−2
   5. $x^4 - 2x^3 + 2x^2 - 2x + 1 = 0$ real roots: 1,1

2. Find the points (real) of intersection of the two curves $f(x) = x^2$ and $g(x) = (2 - 6x)/(x^2 - 3)$.

BUDAN AND STURM METHODS

This final section concerned with finding roots of polynomials presents several methods that are analogous to Descartes's rule of signs in that by counting the number of particular sign changes, you are able to determine the location of zeros. Both methods lend themselves to being programmed because of the nature of the mathematical algorithms involved. In each case you should begin with a polynomial, $p(x) = a_nx^n + a_{n-1}x^{n-1} + \cdots + a_1x + a_0$, with real coefficients. You should consider the equation $p(x) = 0$. You would need to input both the degree n of the polynomial and also the $n+1$ coefficients a_n, a_{n-1}, $\cdots$, a_1, a_0.

If you focus your attention on Budan's method [17] first, you need to generate n additional functions. These functions can be denoted by p_1, p_2, $\cdots$, p_n. These functions are actually what are known as the successive derivatives of p. The function p_{i+1} is the derivative of p_i, so the rule for forming p_1, can then be applied again to form p_2, and then again for p_3, and so on. Let us see, now, how to form p_1.

The function p_1 is a polynomial of degree $(n-1)$ whose coefficients are found by taking the individual coefficients of p and multiplying each by their respective exponent, and then reducing the exponent by one. Thus,

$$p(x) = a_nx^n + a_{n-1}x^{n-1} + a_{n-2}x^{n-2} + \cdots + a_2x^2 + a_1x + a_0$$
$$p_1(x) = na_nx^{n-1} + (n-1)a_{n-1}x^{n-2} + (n-2)a_{n-2}x^{n-3} + \cdots + 2a_2x + 1a_1.$$

To illustrate with a particular example,

$$P(x) = 4x^3 - 2x^2 + 7x - 1$$
$$P_1(x) = 12x^2 - 4x + 7.$$

If the nature of p is maintained in your computer program by the storing of the coefficients in an array, (in the above cubic, the array would be $(4, -2, 7, -1)$. Then p_1 can be computed and stored in an array (one size smaller) by $(4 \cdot 3, -2 \cdot 2, 7 \cdot 1) = (12, -4, 7)$.

Function p_2 is constructed from p_1 by using the same procedure that was used to form p_1 from p. Thus, in general, you have

$$p_2(x) = (n-1)na_n x^{n-2} + (n-2)(n-1)a_{n-1}x^{n-3} + \cdots + (1)2a_2$$

and, in particular, from above,

$$P_2(x) = 24x - 4.$$

Continuing yields

$$p_3(x) = (n-2)(n-1)na_n x^{n-3} + \\ (n-3)(n-2)(n-1)a_{n-1}x^{n-4} + \cdots + (1)(2)(3)a_3$$

and in particular,

$$P_3(x) = 24.$$

It will always follow that the last function generated, p_n, is a constant function, and its value will be $p_n(x) = a_n(n!)$.

The second step in Budan's method is to select two numbers, r and s, such that neither one is a zero of p. Customarily you select consecutive integers, and the polynomial is usually such that it doesn't have integral solutions. You then compute $p(r)$, $p_1(r)$, $p_2(r)$, $\cdots$, $p_n(r)$ and record whether these values are positive or negative. The symbol V_r then represents the number of algebraic sign changes. (Keep in mind that if any $p_i(r)$ happens to equal zero, that term is omitted from sign change consideration. Do you know how to program this?) It follows that the maximum value V_r can assume is n, and the minimum value is zero. Likewise, compute V_s, the number of sign changes in the sequence $p(s)$, $p_1(s)$, $\cdots$, $p_n(s)$.

If you denote the original function p by p_0, the following flow-chart, Fig. 4-5, outlines the computation of V_r.

Let's compute a few values of V_r for various choices of r. You

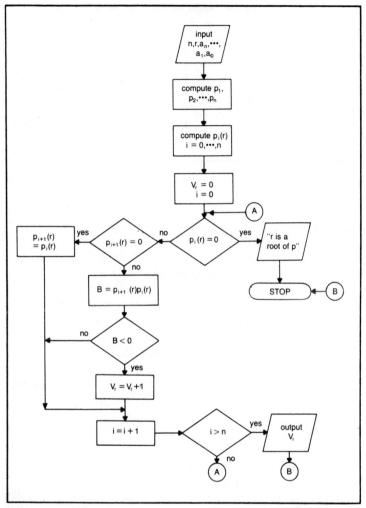

Fig. 4-5. Computation of V_r.

will continue to set $P(x) = 4x^3 - 2x^2 + 7x - 1$. The values are listed in Table 4-4, where instead of listing the exact value of $P_i(r)$, only the algebraic sign is indicated (since that is all that is really important).

Budan's result is that if r, s are not roots to the equation $p(x) = 0$, and if $r < s$, the number of real roots that lie between r and s is either equal to $V_r - V_s$ or is less than that quantity by an even integral amount. See Fig. 4-6.

Table 4-4. Sample Values of $P_i(r)$ and V_r.

r	$P_0(r)$	$P_1(r)$	$P_2(r)$	$P_3(r)$	V_r
−2	−	+	−	+	3
−1	−	+	−	+	3
0	−	+	−	+	3
1	+	+	+	+	0
2	+	+	+	+	0

You therefore know, from the preceding page the example with $P(x) = 4x^3 - 2x^2 + 7x - 1$, that there are no real roots between −2 and 0, and no roots in (1,2). However, the interval (0,1) contains either 3 roots or 3−2 = 1 root. Furthermore, for r > 2, the values of $P(r)$, $P_1(r)$, $P_2(r)$ and $P_3(r)$ are all positive, implying that $V_4 = 0$, so there are no real roots greater than 2. Likewise, for r < −2, it follows that $V_r = 3$, so again there are no real roots to $P(x) = 0$ that are less than −2. All the roots must lie between 0 and 1.

Now let's turn our attention to the result obtained by Sturm. Starting with the function p(x), you need to generate another set of functions, which you can also denote by p_1, p_2, •••, p_m where m ≤ n. The function p_1 is the derivative of p (as in Budan's procedure). The remaining functions are found by a different procedure. You need to perform polynomial division on $p(x)/p_1(x)$, with $p_2(x)$ representing the negative of the remainder. Continuing in this manner, p_{i+1} is the negative of the remainder of the division p_{i-1}/p_i. The process continues until a constant remainder is obtained. In fact, the final remainder, $-p_m$, will be different from zero providing the original polynomial p(x) does not have a repeated factor. These functions p, p_1, p_2, •••, p_m are known as Sturm's functions.

If you set $P(x) = 4x^3 - 2x^2 + 7x - 1$, then $P_1(x) = 12x^2 - 4x + 7$. Since the quotient of $P(x)/P_1(x)$ is x/3 with a remainder of 14x/3 − 1,

$$\frac{P(x)}{P_1(x)} = \frac{x}{3} + \frac{(14/3)x - 1}{P_1(x)}.$$

This means that $P_2(x) = -14x/3 + 1$ (remember, it is the negative of the remainder). To find $P_3(x)$, divide $P_1(x)$ by $P_2(x)$ and get $-36x/14 + 144/196$ with a remainder of $1228/196$.

$$\frac{P_1(x)}{P_2(x)} = \frac{-36x}{14} + \frac{144}{196} + \frac{1228/196}{P_2(x)}$$

Thus, $P_3(x) = -1228/196 \doteq -6.27$.

You will have to program the determination of these polynomials. But what this really entails is that you must be able to divide two polynomials in which the degree of the denominator polynomial is no greater than that of the numerator. Of course all computations will be done with decimals, and you will want to keep track of the remainder.

To compute the Sturm functions by hand can be quite laborious, especially if p is of high degree. The real labor is in keeping all the coefficients in exact fraction form. The computer introduces some roundoff error by converting these coefficients to decimals, but for the most part, the final results will still be good.

If you let r denote any real number that is not a solution to $p(x) = 0$ the symbol v_r will denote the number of variations in sign of the $(m + 1)$ values,

$$p(r),\ p_1(r),\ p_2(r),\ \bullet\bullet\bullet,\ p_m(r).$$

The results shown in Table 4-5 demonstrate various values of v_r for the function $p(x) = 4x^3 - 2x^2 + 7x - 1$.

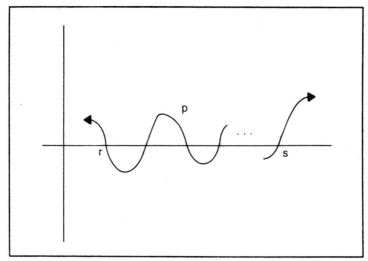

Fig. 4-6. Number of roots in (r,s) equals $V_r - V_s - 2k$.

Table 4-5. Various Values of V_r.

r	p(r)	$p_1(r)$	$p_2(r)$	$p_3(r)$	V_r
−2	−55	63	31/3	−6.27	2
−1	−14	23	17/3	−6.27	2
0	−1	7	1	−6.27	2
1	8	15	−11/3	−6.27	1
2	37	47	−25/3	−6.27	1

The numerical values in the table under the column headings p(r), $p_1(r)$, $p_2(r)$, and $p_3(r)$ are not important; only their algebraic sign is. Sturm's theorem can be stated now. You should note its similarity to Budan's result and its difference from that result. Sturm's result says that if p(x) = 0 has no multiple roots, and if neither r nor s are roots of the equation, then the number of roots that lie between r and s is exactly equal to $v_r - v_s$. Consequently, the equation $4x^3 - 2x^2 + 7x - 1 = 0$ has precisely one root between 0 and 1, assuming it has no multiple roots. You know there are no roots between 1 and 2, and no roots between −2 and 0.

I will give one final example that demonstrates how to isolate real roots by employing the conclusions of both Budan and Sturm.

Example 1. Let $p(x) = 3x^4 - 24x^2 + 64x - 48$. Using Budan's results you have:

$$p_1(x) = 12x^3 - 48x + 64$$
$$p_2(x) = 36x^2 - 48$$
$$p_3(x) = 72x$$
$$p_4(x) = 72$$

Setting r = −4, −3, •••, 2, 3 gives the results shown in Table 4-6. From this you can deduce that there is one root in the interval (−4,−3), one root in (0,1), and either two roots, or no roots, in (1,2).

On the other hand, the Sturm functions are

$$p_1(x) = 12x^3 - 48x + 64$$
$$p_2(x) = 12x^2 - 48x + 48$$
$$p_3(x) = -96x + 128$$
$$p_4(x) = -16/3$$

Table 4-6. Computing V_r when $p(x) = 3x^4 - 24x^2 + 64x - 48$.

(r)	p(r)	$p_1(r)$	$p_2(r)$	$p_3(r)$	$p_4(r)$	V_r
-4	+	-	+	-	+	4
-3	-	-	+	-	+	3
-2	-	.+	+	-	+	3
-1	-	+	-	-	+	3
0	-	+	-		+	3
1	+	+	-	+	+	2
2	+	+	+	+	+	0
3	+	+	+	+	+	0

and the analogous list of sign variations is given in Table 4-7.

You know there are no roots in (1,2). Furthermore, $v_r = 1$ if $r > 3$, and $v_r = 3$ for $r < -4$: this implies the equation has two imaginary roots in addition to the two real roots. In both cases note that $p(r)$ is never equal to zero, which is a condition that must be fulfilled.

Table 4-7. Computing v_r when $p(x) = 3x^4 - 24x^2 + 64x - 48$.

r	p(r)	$p_1(r)$	$p_2(r)$	$p_3(r)$	$p_4(r)$	V_r
-4	+	-	+	+	-	3
-3	-	-	+	+	-	2
-2	-	+	+	+	-	2
-1	-	+	+	+	-	2
0	-	+	+	+	-	2
1	+	+	+	+	-	1
2	+	+	+	-	-	1
3	+	+	+	-	-	1

Exercises

1. Write a program that computes V_r and v_r for $p(x) = 0$. This means you must use both Budan's method and Sturm's method. Test your program with the following polynomials.

a. $p(x) = x^3 - 12x^2 + 48x + 3$ one real root in $(-1,0)$

b. $p(x) = x^4 - 4x^3 + 7x^2 - 2x + 5$ no real roots

c. $p(x) = x^5 + 2x^4 + x^3 - 4x^2 - 3x - 5$ one real root in the interval $(1,2)$.

153

In all cases, let r run through the integral values -10, -9, $\cdots$, 8, 9, 10.

2. Prove that the equation $x^5 - x^4 - 21x^3 - 5x^2 + 53x + 51 = 0$ has no root greater than 5.

CONICS

It is customary for students to be exposed to the study of conics sometime in high school. *Conics* typically means the curves described by a circle, an ellipse, a parabola, or a hyperbola as shown in Fig. 4-7.

In plane geometry these four curves are presented and defined by the following:

A *circle* is described as the locus of all points in the plane that are equidistant from a given point (the center of the circle).

An *ellipse* is described as the locus of all points in the plane whose

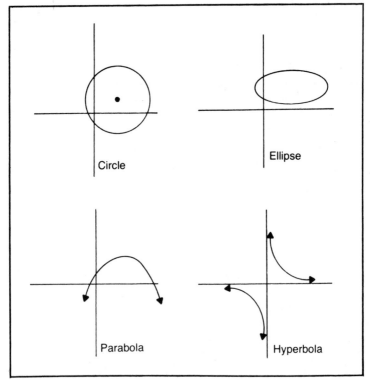

Fig. 4-7. The conics.

154

distances from two given points (the foci) sums to a given positive constant.

A *parabola* is described as the locus of all points in the plane that are equidistant from a given point (the focus) and a given line (the directrix).

A *hyperbola* is described as the locus of all points in the plane whose distances from two given points (the foci) differs by a given constant.

These are precisely the curves generated when a plane intersects a right-circular cone of two nappes, hence the origin of the phrase *comic sections*. The Greek mathematician Apollonius studied the conics in terms of geometry by using this concept. But for the remainder of this section, the conics will be studied based on their analytical definitions using the algebraic equations that so completely characterize the curves.

The section will concentrate on those equations of the form

$$Ax^2 + Bxy + Cy^2 + Dx + Ey + F = 0,$$

where the coefficients are real constants. These equations give rise to the four conics (circle, ellipse, parabola, hyperbola), when the coefficients are related in the appropriate way. These are the most interesting graphs, although other curves, known as degenerate conics, could be produced under the right set of circumstances. Degenerate conics include a single line, several lines, a single point, or the empty set. Examples of how these degenerates, along with the four standard conics, could be produced from one of the above quadratic equations, are given in Table 4-8.

Now let us return to the general equation

$$Ax^2 + Bxy + Cy^2 + Dx + Ey + F = 0$$

and see what conditions A, B, C, D, E, and F must meet in order for a conic to be produced. You should note that after these six values are fed in as input, a series of decision statements will be needed to help determine the exact nature of the conic. This discussion will be divided into two main classes depending on whether B is zero or nonzero.

Case 1: B = 0
    Subcase 1: If A = C = 0, the equation reduces to a linear equation, $Dx + Ey + F = 0$, or equivalently,

155

Table 4-8. Conics and Degenerate Conics.

Curve	Equation
empty set	$x^2 + 3y^2 + 4 = 0$
single point	$x^2 + 3y^2 = 0$
line	$2x + 3y + 4 = 0$
two intersecting lines	$4x^2 - y^2 = 0$
two parallel lines	$x^2 - 9x = 0$
circle	$x^2 + y^2 + x + 2y + 1 = 0$
ellipse	$x^2 + 2y^2 + x + 8y + 1 = 0$
parabola	$x^2 + 2xy + y^2 + x + y = 0$
hyperbola	$x^2 + 8xy + y^2 + x = 0$

$$y = -\frac{D}{E} x - \frac{F}{E}$$

where the slope of the line is $-D/E$, and the y-intercept is $-F/E$. Of course, you may wish to consider the cases in which $D = 0$, or $E = 0$. These yield horizontal and vertical lines. You may also wish to try other combinations in which these coefficients vanish.

Subcase 2: If $A = 0$ and $C \neq 0$, the graph is either a parabola or a degenerate parabola, which in this case would include either two parallel lines, one line, or the empty set. The degenerate case would occur if $D = 0$; otherwise the curve is a parabola of the form

$$(y - k)^2 = 4p(x - h).$$

In this form, h, p, and k would be represented by

$$k = \frac{-E}{2C}$$

$$p = \frac{-D}{4C}$$

$$h = \frac{E^2 - 4CF}{4CD}$$

The importance of h, k, and p is that (h,k) represents the vertex of the parabola, and the focus and directrix are located p units from the vertex. Thus, the line $x = h - p$ is the directrix, and (h + p,k) is the focus. If you consider the particular equation

$$3y^2 + 2x + 6y + 4 = 0,$$

you find that $k = -6/(2 \cdot 3) = -1$, $p = -2/(4 \cdot 3) = -1/6$, and $h = -1/2$. This gives you the graph pictured in Fig. 4-8, where the

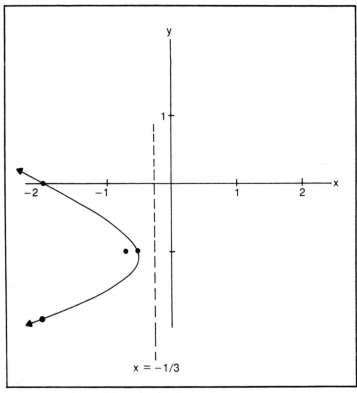

Fig. 4-8. The parabola $3y^2 + 2x + 6y + 4 = 0$.

vertex is at $(-1/2,-1)$, the focus at $(-2/3, -1)$, and the directrix is $x = -1/3$.

Subcase 3: If $A \neq 0$ and $C = 0$, you are in the same situation as in subcase (2), except the variables x and y are interchanged. The graph is either a parabola (if $D \neq 0$), or is degenerate (if $D = 0$). If it is not degenerate, the parabola is of the form

$$(x - h)^2 = 4p(y - k)$$

where (h,k) is the vertex, and the focus $F(h,k + p)$ and directrix $y = k - p$ are p units away from the vertex. In this case, note that the directrix is a horizontal line, while it was a vertical line in the previous case. Furthermore, the values of h, k, and p are given by

$$h = \frac{-D}{2A}$$

$$p = \frac{-E}{4A}$$

$$k = \frac{D^2 - 4AF}{4AE} \quad .$$

The particular equation $x^2 + 6x - 12y + 57 = 0$ graphs as a parabola with vertex at $(-3, 4)$, and since $p = 3$, the focus is $F(-3,7)$, and $y = 1$ is the directrix. See Fig. 4-9.

Subcase 4: If $A \neq 0$, $C \neq 0$, and $AC > 0$, the graph is either an ellipse or a degenerate ellipse. Assume that both A and C are positive, since otherwise you could multiply every coefficient by -1 and produce an equivalent equation. Then note that the equation

$$Ax^2 + Cy^2 + Dx + Ey + F = 0$$

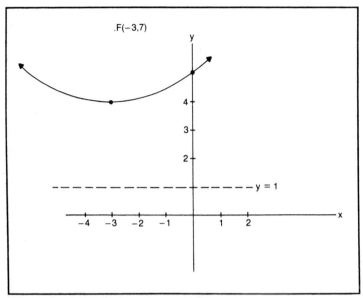

Fig. 4-9. The parabola $x^2 + 6x - 12y + 57 = 0$.

can be written as

$$A\left(x^2 + \frac{Dx}{A}\right) + C\left(y^2 + \frac{Ey}{C}\right) = -F.$$

By completing the square on both binomials, you get

$$A\left(x + \frac{D}{2A}\right)^2 + C\left(y + \frac{E}{2C}\right)^2 = -F + \frac{D^2}{4A} + \frac{E^2}{4C}.$$

Since the left-hand side of this equation is nonnegative, if the right-hand side is negative, or

$$\frac{D^2}{4A} + \frac{E^2}{4C} < F$$

the graph is degenerate because the equation has no solution. If the right-hand side equals zero, the graph is also degenerate because

159

the only solution is the single point, x = −D/(2A), y = −E/(2C). If the right-hand side is positive, you get a legitimate ellipse of the form

$$\frac{(x - h)^2}{r^2} + \frac{(y - k)^2}{s^2} = 1.$$

In this standard form, h would be equal to −D/(2A); k would be equal to −E/(2C); r^2 would equal $D^2/(4A^2) + E^2/(4AC) - F/A$; and $s^2 = D^2/(4AC) + E^2/(4C^2) - F/C$. The point (h,k) represents the center of the ellipse; r represents the length of the horizontal axis; and s the length of the vertical axis. These lengths are shown in Fig. 4-10. It is important to know which axis is the principal axis (the longer one) because the two foci are located on this axis. If r > s, the principal (major) axis is parallel to the x-axis, and the two foci are

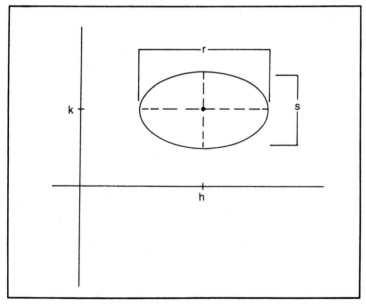

Fig. 4-10. An ellipse.

located $\sqrt{r^2-s^2}$ units from the center. If $s > r$, the major axis is parallel to the y-axis, and the foci are $\sqrt{s^2-r^2}$ units from the center. Consequently, in this latter case, with $s > r$, the coordinates of the foci would be

$$F(h, k \pm \sqrt{s^2-r^2}).$$

To illustrate subcase 4, note that the equation

$$6x^2 + 9y^2 - 24x - 54y + 115 = 0$$

has no solution since it can be rearranged as

$$6(x - 2)^2 + 9(y - 3)^2 = -10.$$

On the other hand,

$$6x^2 + 9y^2 - 24x - 54y + 51 = 0$$

can be rearranged and put in standard form as

$$\frac{(x - 2)^2}{9} + \frac{(y - 3)^2}{6} = 1.$$

Thus $r = 3$ and $s = \sqrt{6}$, and the ellipse has its center at $(2,3)$, with the major axis running horizontal and the foci located at $(2 \pm \sqrt{3}, 3)$.

Subcase 5: If $A \neq 0, C \neq 0$, and $AC < 0$, the graph is either a hyperbola or a degenerate hyperbola, which in this case would amount to two intersecting straight lines. By rearranging terms and completing the square the main equation can read

$$A\left(x + \frac{D}{2A}\right)^2 + C\left(y + \frac{E}{2C}\right)^2 = \frac{D^2C + E^2A - 4ACF}{4AC}.$$

If the term on the right, call it T, equals zero (thus $D^2C + E^2A = 4ACF$), the curve will degenerate into the intersecting lines, namely

$$\sqrt{|A|}\left(x + \frac{D}{2A}\right) + \sqrt{|C|}\left(y + \frac{E}{2C}\right) = 0.$$

But if the term on the right is positive, and if A is positive (so C must be negative), the hyperbola opens to the right and left as shown in Fig. 4-11, and is of the standard form

$$\frac{(x - h)^2}{r^2} - \frac{(y - k)^2}{s^2} = 1$$

where $h = -D/(2A)$, $k = -E/(2C)$, $r^2 = (D^2C + E^2A - 4ACF)/(4A^2C)$, and $s^2 = (4ACF - D^2C - E^2A)/(4AC^2)$. The point (h,k) represents the center of the hyperbola. The two foci, F_1 and F_2, are located along the horizontal line $y = k$ at a distance of $\sqrt{r^2 + s^2}$ from the center. Thus, you have $F_1(h + \sqrt{r^2 + s^2}, k)$

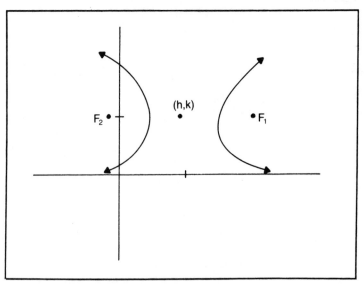

Fig. 4-11. Hyperbola opening right and left.

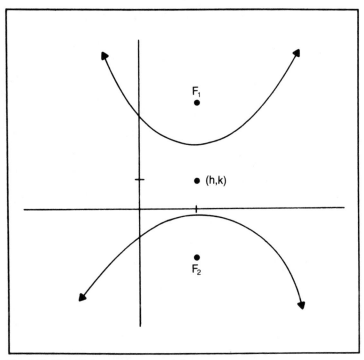

Fig. 4-12. Hyperbola opening up and down.

and $F_2(h - \sqrt{r^2 + s^2}, k)$. This situation would also happen if $A < 0$ and the term T were negative. Otherwise, if $AT < 0$ the hyperbola would open up and down, as in Fig. 4-12. The standard form for this hyperbola would be

$$\frac{(y - k)^2}{r^2} - \frac{(x - h)^2}{s^2} = 1$$

where $h = -D/(2A)$, $k = -E/(2C)$, $r^2 = (D^2C + E^2A - 4ACF)/(4AC^2)$, and $s^2 = (4ACF - D^2C - E^2A)/4A^2C)$. Again, (h,k) is the center of the hyperbola, but F_1 and F_2 lie along the vertical line $x = h$ at a distance of $\sqrt{r^2 + s^2}$ from the center. To illustrate, the hyperbola

$$-4x^2 + 3y^2 - 8x - 24y - 40 = 0$$

163

can be put in standard form as

$$\frac{(y-4)^2}{28} - \frac{(x+1)^2}{21} = 1$$

so the center is at $(-1,4)$. This hyperbola opens up and down, and the foci are located $\sqrt{28+21} = 7$ units from the center, at $F_1(-1,11)$ and $F_2(-1,-3)$.

Case II: $B \neq 0$

What happens in this case is that one of the conics has been disguised because the x-y axes have been rotated through an angle of θ radians ($-\pi/2 \leq \theta \leq \pi/2$). The exact angle of rotation can be determined from the equation

$$\cot (2\theta) = \frac{A-C}{B}.$$

The equation $17x^2 - 12xy + 8y^2 = 20$ represents an ellipse that has been rotated through an angle of approximately 1.1073 radians ($\cot(2\theta) = -3/4$), as shown in Fig. 4-13. Some of the calculations get a little messy when

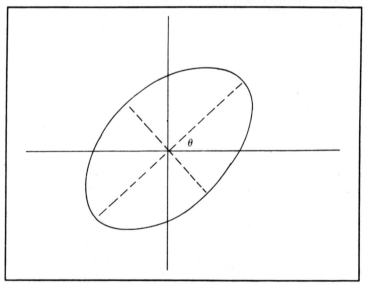

Fig. 4-13. Ellipse rotated through θ radians.

axes are rotated, so it is sufficient to say that the type of conic produced can be determined by the algebraic sign of the discriminant $B^2 - 4AC$. This gives you

   a. a parabola or a degenerate parabola if $B^2 - 4AC = 0$

   b. an ellipse or a degenerate ellipse if $B^2 - 4AC < 0$

   c. a hyperbola or a degenerate hyperbola if $B^2 - 4AC > 0$.

You are referred to any standard calculus text for further information on this subject.

Exercises

1. Write a program that determines what kind of conic is produced from the equation $Ax^2 + Bxy + Cy^2 + Dx + Ey + F = 0$ when various values for the six coefficients are fed into the program. If $B \neq 0$, just give the nature of the conic and approximate the angle of axis rotation. Otherwise, if $B = 0$, be explicit in describing the nature. Output the center and coordinates of the foci. If the curve is degenerate, state its nature and characteristic property (i.e., if it is a line, give the equation; if it is a point, give the coordinates). Test your program with the following equations.

    1. $x^2 + 2y^2 = 0$
    2. $6x^2 - y^2 = 0$
    3. $x^2 + y^2 + 2x + 4y + 1 = 0$
    4. $2x^2 + y^2 + 3x + 2y + 1 = 0$
    5. $2x^2 + 3x + 4y - 5 = 0$
    6. $x^2 + 8xy + y^2 + x - 1 = 0$.

2. The ceiling in a 20 ft wide hallway is in the shape of a semi-ellipse. The ceiling is 18 ft high in the center and 12 ft high at the side walls. Find the height of the ceiling 4 ft from either wall.

COUNTABILITY OF RATIONAL POLYNOMIALS

When it comes to talking about the infinite, that mysterious abstraction created by man, only those well versed in the matter should participate. The infinite is clearly a topic that very few understand, yet nearly everyone has a working knowledge and general feel for it. You wouldn't want to say, "there are an infinite number of grains of sand along the shore of Lake Erie," because it wouldn't be true. But if you did say it, chances are your companion wouldn't correct you—that is, unless your companion was a mathematician.

It is somewhat difficult, at least at first glance, to demonstrate infinite sets of tangible items here in the real world. There aren't infinitely many people or infinitely many books or infinitely many grains of sand. How about infinitely many stars? Well, no one knows for sure. What sets are infinite, if none of these are? To answer that, consider all of the following: a bullet shot from a gun may land in any one of an infinite number of places; a light striking a pole produces a shadow with an infinite number of different possible angles of elevation; a plant will grow to any one of an infinite number of different heights; and, to satisfy man's mental appetite, there are an infinite number of both positive integral multiples of 17 and non-differentiable functions defined on the unit interval.

So much for some elementary examples. When mathematicians discuss the infinite, they have to be careful because they often need to distinguish between different levels of infinity. Yes, that's right; there are different levels, or classifications, of the infinite. In fact, there are infinitely many different levels! Let us see if we can't give an intuitive explanation of how this can be.

To begin with, you have the set of natural numbers N, with

$$N = \{1, 2, 3, \cdots\}$$

where N is sometimes called the set of counting numbers. Suppose you wish to count how many items there are in a particular collection. This means there will be a first item, a second item, a third item, and so on until you reach the last item, say the kth item. You then say the collection contains k items. One way to interpret this counting process is to view the situation as a pairing-up of items of the set with some initial subset of N. The subset in this case was $\{1, 2, 3, \cdots, k\}$. The diagram in Fig. 4-14 might help to explain this.

Each element of the set is paired-up with a different natural number, and each natural number 1, 2, $\cdots$, k is paired with some unique member of the set. The elements of the set are said to be in one-to-one correspondence with the subset $\{1, 2, \cdots, k\}$. When this happens, the largest natural number in the subset (in this case k) is the count of how many elements are in the set.

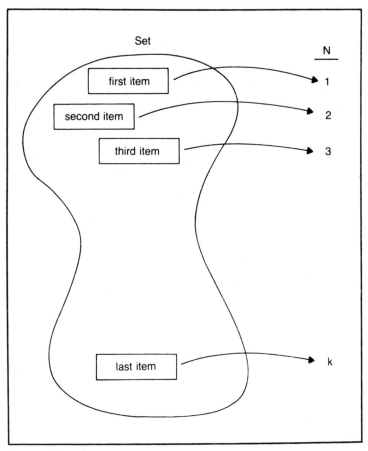

Fig. 4-14. Counting.

For a second example, if you wished to count the number of fingers on your left hand, you could set up a one-to-one correspondence as shown in Fig. 4-15. The fingers are paired up with the numbers 1, 2, 3, 4, 5, so 5, which is the largest number, indicates how many fingers you have.

The aforementioned set of positive integral multiples of 17 presents a new problem. If you set up the correspondence with N as shown in Fig. 4-16, you see that there is no largest value of k to indicate how many elements the set contains. This means the set is infinite! Any set that can not be placed in a one-to-one correspondence with some initial subset $\{1, 2, \bullet\bullet\bullet, k\}$ of N is defined to be infinite.

167

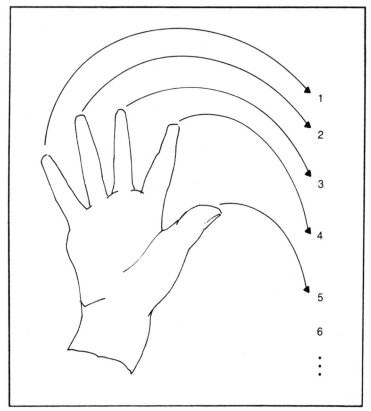

Fig. 4-15. Counting fingers.

Now comes the difficult part. If a set is infinite, and if it can be paired up in a one-to-one fashion with N, the set is said to be countably infinite. This is the lowest of all levels of infinite. Examples of sets that are countably infinite include.

1. {positive multiples of 10} = {10, 20, 30, •••}
2. {even numbers} = {0, 2, 4, 6, •••}
3. {negative integers} = {−1, −2, −3, •••}
4. {positive rational numbers} = {1, 1/2, 2, 3, 1/3, 1/4, 2/3, •••}.

Infinite sets that are not countably infinite are said to be uncountable (and there are infinitely many different levels of uncountability), and some examples would include

1. {real numbers between 0 and 1} = (0,1)
2. {different elevation angles for a shadow}
3. {positive numbers} = (0, ∞)
4. {different heights for a particular plant}
5. {irrational numbers between 0 and 3}.

It is usually not too difficult (although sometimes it is a monumental task) to prove that some particular set is infinite, but it is a harder job to determine whether the set is countably infinite or uncountable.

Consider the set Q of positive rational numbers

$$Q = \{m/n \mid m, n \text{ are positive integers}\}$$

and see if you can show why it is countable. If you write the elements of Q in an infinite rectangular array, where the i-th row contains all the positive rational numbers, in ascending order, with a denominator of i, then you know that every element of Q is listed somewhere in this array. You then start counting the elements in this array, by choosing 1 first, then following along the path indicated in Fig. 4-17. Numbers that are equal, such as $1 = 2/2 = 3/3 = \cdots$, or $1/2 = 2/4 = 3/6 = \cdots$, are not to be counted more than once. This means the elements of Q can be placed in a one-to-one correspondence with N as shown in Fig. 4-18, so Q is countably infinite.

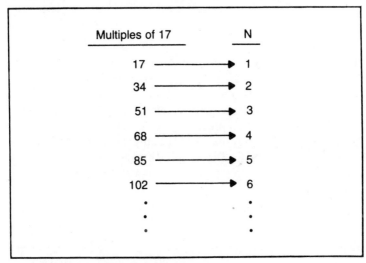

Fig. 4-16. Counting the multiples of 17.

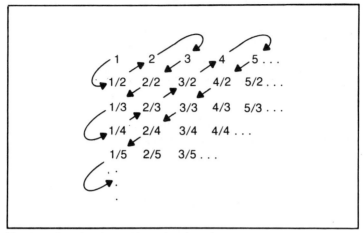

Fig. 4-17. Counting the positive rational numbers.

This method of counting Q is known as Cantor's diagonalization process.

Closely related to Q is the set Q_p, which is the set that will be dealt with in the remainder of this section. This is the set of all polynomials with rational (positive or negative) coefficients. Thus, Q_p would contain at least the following polynomials:

1. $2x^2 - (3/2)x + 1$
2. 4.7
3. $(1/3)x - 7$
4. $10x^{18} - (2/5)x.$

It is quite obvious that Q_p is an infinite set, but is it countably infinite or uncountable? According to [26], the set is countably infinite, and in fact, [26] presents a nice algorithm for computing the natural number that is paired up with a given polynomial $p(x)$ from Q_p.

To understand this process, you will have to input into your program approximately the first dozen primes: $PRIME(1) = 2$, $PRIME(2) = 3$, $PRIME(3) = 5$, $PRIME(4) = 7$, •••, $PRIME(12) = 37$. You can begin with a typical polynomial $p(x)$ of the form

$$p(x) = a_n x^n + a_{n-1} x^{n-1} + \cdots + a_1 x + a_0$$

where all the a_i are rational numbers and $a_n \neq 0$. This means each a_i has a numerator, NUMER(i), and a denominator, DENOM(i).

$$a_i = \frac{NUMER(i)}{DENOM(i)}$$

These values will also have to be fed into the program as input. If some $a_i = 0$, it suffices to set $NUMER(i) = 0$ and $DENOM(i) = 1$.

Next, set LCD to stand for the least common denominator of all the $(n + 1)$ coefficients of $p(x)$. Then note that $p(x)$ can be expressed in standard form as

$$p(x) = \frac{1}{LCD} \left[(-1)^{s_n} b_n x^n + (-1)^{s_{n-1}} b_{n-1} x^{n-1} + \cdots + (-1)^{s_0} b_0 \right]$$

where the b_i are all nonnegative integers, each s_i is zero or one, and

$$\frac{(-1)^{s_i} b_i}{LCD} = a_i .$$

To illustrate, consider the specific polynomial $p(x) = (1/4x^2 - x + 2/3$. You have $n = 2$, $a_2 = 1/4$, $a_1 = -1$, and $a_0 = 2/3$. It follows that $LCD = 12$, and

Q	N
1	1
1/2	2
2	3
3	4
1/3	5
1/4	6
2/3	7
•	•
•	•
•	•

Fig. 4-18. The correspondence of Q with N.

$$a_2 = \frac{1}{4} = \frac{3}{12} = \frac{1}{12}(-1)^0(3)$$

$$a_1 = -1 = \frac{-12}{12} = \frac{1}{12}(-1)^1(12)$$

$$a_0 = \frac{2}{3} = \frac{8}{12} = \frac{1}{12}(-1)^0(8)$$

so $s_2 = 0$, $s_1 = 1$, $s_0 = 0$ and $b_2 = 3$, $b_1 = 12$, $b_0 = 8$. Thus, $p(x)$ is rewritten as

$$p(x) = \frac{1}{12}[(-1)^0\, 3x^2 + (-1)^1\, 12x + (-1)^0 8].$$

Can you see what you would have to do in your program to put $p(x)$ in standard form? The major task would be in computing LCD.

Once $p(x)$ is in standard form, the exponents s_i are used to form a unique number, call it SUM, using the relationship

$$\text{SUM} = s_n 2^n + s_{n-1}\, 2^{n-1} + \cdots + s_1 2 + s_0.$$

This value SUM can be easily computed: the following four lines of BASIC will suffice.

```
10   SUM = 0
20   FOR I = 0 TO N
30        SUM = SUM + (2**I)*S(I)
40   NEXT I
```

Finally, the natural number, NUMBER, that is paired with $p(x)$ is computed as the following product.

$$\text{NUMBER} = \text{PRIME}(1)^{\text{SUM}}\text{PRIME}(2)^{\text{LCD}}\text{PRIME}(3)^{b_0}\text{PRIME}(4)^{b_1}$$
$$\cdots \text{PRIME}(n+3)^{b_n}.$$

Therefore, to finish off the example with $p(x) = (1/4x)^2 - x + 2/3$, you have

$$\text{LCD} = 12$$
$$b_0 = 8$$

$$b_1 = 12$$
$$b_2 = 3$$
$$\text{SUM} = 0 \cdot 2^2 + 1 \cdot 2^1 + 0 \cdot 2^0 = 2$$

so

$$\text{NUMBER} = 2^2 \cdot 3^{12} \cdot 5^8 \cdot 7^{12} \cdot 11^3.$$

It would be best to leave the number in this product form since it is so large. You should be aware of the fact that this number, even if multiplied out, could only be arrived at using the special combination of primes and exponents shown. No other combination of integers could produce NUMBER, and thus no other polynomial could be paired with NUMBER. Conversely, given any positive integer, its prime factorization determines SUM, LCD, n, and b_0, b_1, $\cdots$, b_n. Thus, the polynomial $p(x)$ is uniquely determined. What does all of this mean? It means Q_p is countably infinite!

Exercises

1. Suppose $n \leq 10$. Write a program that computes NUMBER (leave it in factored form) for a polynomial $p(x)$ of the form

$$p(x) = a_n x^n + a_{n-1} x^{n-1} + \cdots + a_0$$

with rational coefficients. Test your program with the following polynomials.

    1. $p(x) = x^2/4 - x + 2/3$
    2. $p(x) = 3x^5 + (2/5)x^3 - x^2/7 + .1x - 2.$

2. What polynomial $p(x)$ does NUMBER $= 9,507,960$ correspond to?

PIVOTING

    Quite often a physical situation involves a number of variables, say x_1, x_2, $\cdots$, x_n, and an equal number of constraints binding these variables together in linear form. This produces a system of n linear equations in n unknowns:

$$a_{11}x_1 + a_{12}x_2 + \cdots + a_{1n}x_n = b_1$$
$$a_{21}x_1 + a_{22}x_2 + \cdots + a_{2n}x_n = b_2$$

$$\vdots$$

$$a_{n1}x_1 + a_{n2}x_2 + \cdots + a_{nn}x_n = b_n.$$

This system can be written in a more mathematically compact form as

$$\sum_{j=1}^{n} a_{ij}x_j = b_i$$

where i runs through all the values 1, 2, $\cdots$, n. It is to be understood that the coefficients a_n and b_i all represent real numbers, numbers that are known and given from the content of the problem. These numbers are all fed into your program as input data. The variables i, j are subscripts, and are used to determine the correct location of a_{ij}. The variable i denotes the row equation, and j denotes the appropriate column, which means it is the coefficient of the variable x_j. For programming purposes, the coefficient a_{ij} would probably be written as an array A(i,j).

The specific system

$$\begin{aligned} 2x_1 + 3x_2 - x_3 &= 6 \\ x_1 + 2x_2 + x_3 &= 5 \\ -x_1 - x_2 + 4x_3 &= 1 \end{aligned}$$

consists of 3 equations in 3 unknowns, x_1, x_2, and x_3. The coefficients are $a_{11} = 2$, $a_{12} = 3$, $a_{13} = -1$, $a_{21} = 1$, $a_{22} = 2$, $a_{23} = 1$, $a_{31} = -1$, $a_{32} = -1$, and $a_{33} = 4$. Furthermore, $b_1 = 6$, $b_2 = 5$, and $b_3 = 1$. This particular system has the unique solution $x_1 = 2$, $x_2 = 1$ and $x_3 = 1$.

It is left to mathematics to devise a procedure for solving a given system of linear equations. One of the first methods students learn in high school algebra is that of substitution. In this, you solve for one of the variables, say x_1, in one of the equations: ideally one in which the coefficient of x_1 is one. This expression is then substituted for x_1 in all the other equations. This produces a new system of $(n-1)$ equations in $(n-1)$ unknowns. You then solve for x_2 (in terms of x_3, x_4, $\cdots$, x_n) in one of these equations, and this expression is substituted for x_2 in the remaining $(n-2)$ equations; the process continues with an expression for x_{n-1} substituted into the nth and last equation. This gives a simple linear equation in one variable, x_n, which can be solved. Once x_n is known, you backtrack and compute $x_{n-1}, \bullet\bullet\bullet, x_2$ and x_1.

As an example, solve for x_1 from the third equation above. First you will get

$$x_1 = -x_2 + 4x_3 - 1.$$

Substituting this in the first two equations gives

$$x_2 + 7x_3 = 8$$
$$x_2 + 5x_3 = 6.$$

Solving for x_2 in the second equation ($x_2 = -5x_3 + 6$) and substituting the result the first equation gives $2x_3 = 2$, or $x_3 = 1$. It follows that $x_2 = -5(1) + 6 = 1$ and $x_1 = -1 + 4(1) - 1 = 2$.

The substitution method works well when you are solving the system by hand, especially if n isn't too large. It's a difficult method to program, so programmers use a method termed *Gaussian elimination* (named after Karl Frederick Gauss). To illustrate this method (also familiar to all high school students), consider the system

$$2x_1 + 9x_2 - 3x_3 = 4$$
$$4x_1 - 3x_2 + x_3 = 1$$
$$-3x_1 + x_2 + x_3 = 1/6.$$

The first step is to reduce this system to an equivalent one with two equations in two unknowns, say x_2 and x_3. These are equivalent in the sense that both systems will have the same solution values for x_2 and x_3. Thus, to eliminate x_1, you multiply the first equation by $-4/2$ and add this equation to the second equation, giving $-21x_2 + 7x_3 = -7$. You then multiply the first equation by $-(3)/2$ and add to the third equation, giving $(29/2)x_2 - (7/2)x_3 = 37/6$. These last two equations form the new system of two equations in two unknowns. The second step is to reduce this system to one equation in one unknown. To this end, you multiply the first equation in the reduced system by $(-29/2)/(-21)$ and add to the second, giving $(8/6)x_3 = 8/6$, from which $x_3 = 1$. It then follows that $x_2 = 2/3$ and $x_1 = 1/2$.

The advantage of the Gaussian elimination method is that the method can be easily programmed: the same sequence of steps is applied repeatedly. Note that when given the original system of equations,

$$a_{11}x_1 + a_{12}x_2 + \cdots + a_{1n}x_n = b_1$$
$$a_{21}x_1 + a_{22}x_2 + \cdots + a_{2n}x_n = b_2$$
$$\bullet$$
$$\bullet$$
$$\bullet$$
$$a_{n1}x_1 + a_{n2}x_2 + \cdots + a_{nn}x_n = b_n$$

if you multiply the first equation by $-a_{k1}/a_{11}$, for each $k = 2,3,\cdots,n$, and add that to the equation in the k-th row, a new system will be produced, consisting of $(n-1)$ equations in $(n-1)$ unknowns. The procedure is repeated, so that in the next step, you eliminate the variable x_2, and then x_3, and so on.

The disadvantage of the Gaussian method is somewhat subtle, yet significant. Each of the $(n-1)$ divisions that take place when you are eliminating x_1 from the system constitutes a source for roundoff error. This error is pronounced when a_{11} is very small (and of course you are in much trouble when $a_{11} = 0$), and especially so if a_{11} is smaller than the computer will accurately handle. The same problem could develop during any of the $(n-2)$ divisions that take place when you are eliminating x_2 from the system.

Fortunately, a lot of these problems can be eliminated if you interchange the position of the equations at the appropriate time and in the appropriate way so as to minimize roundoff error incurred during all the $(n-1) + (n-2) + \cdots + 2 + 1$ divisions. This is the principle behind the Gaussian elimination with pivoting [15].

Suppose you wish to solve the system

$$3x_1 + x_2 + 7x_3 = -2$$
$$-2x_1 + x_2 + 3x_3 = -3$$
$$x_1 + 4x_2 - x_3 = 10.$$

You might find it more useful to consider this as a 3×4 matrix array A, where

$$A = \begin{bmatrix} 3 & 1 & 7 & -2 \\ -2 & 1 & 3 & -3 \\ 1 & 4 & -1 & 10 \end{bmatrix}.$$

First focus your attention on those 9 numbers that are the coefficients of x_1, x_2, and x_3; i.e., A(i,j), $i \le 3$, $j \le 3$. Then $s_{i,1}$ is defined to be the maximum of the absolute value of the three coefficients from the i-th row of A, with i = 1, 2, 3:

$$s_{i,1} = \max_{1 \le j \le 3} |A(i,j)|.$$

Thus, $s_{1,1} = 7$, $s_{2,1} = 3$ and $s_{3,1} = 4$. You then form the three ratios $|A(i,1)|/s_{i,1}$ and see which has the largest value. These three ratios are 3/7, 2/3, and 1/4, so 2/3 is the largest: this means that the second row in A is your pivot row.

$$A = \begin{bmatrix} 3 & 1 & 7 & -2 \\ -2 & 1 & 3 & -3 \\ 1 & 4 & -1 & 10 \end{bmatrix} \text{ pivot row.}$$

This pivot row is interchanged with the first row of A, producing a new array, which is still labeled A,

$$A = \begin{bmatrix} -2 & 1 & 3 & -3 \\ 3 & 1 & 7 & -2 \\ 1 & 4 & -1 & 10 \end{bmatrix} .$$

You should realize that you are now trying to solve the system of equations,

$$\begin{aligned} -2x_1 + x_2 + 3x_3 &= -3 \\ 3x_1 + x_2 + 7x_3 &= -2 \\ x_1 + 4x_2 - x_3 &= 10 \end{aligned}$$

which is certainly equivalent to the initial system.

Gaussian elimination is then performed on A: you multiply row 1 by $-3/-2$ and add to the second row. You multiply row 1 by $-1/-2$ and add to the third row. This gives a new array, also called A,

$$A = \begin{bmatrix} -2 & 1 & 3 & -3 \\ 0 & 2.5 & 11.5 & -6.5 \\ 0 & 4.5 & .5 & 8.5 \end{bmatrix} .$$

You can then define $s_{i,2}$ to be the maximum of the absolute value of the two coefficients of x_2 and x_3 from the i-th row of A, with $i = 2, 3$:

$$s_{i,2} = \max_{2 \leqslant j \leqslant 3} |A(i,j)|.$$

Thus, $s_{2,2} = 11.5$ and $s_{3,2} = 4.5$. Again, you must seek the largest of the ratios $|A(i,2)|/s_{i,2}$ because this will determine your pivot row. Since the two ratios are $2.5/11.5 = 2/11$ and $4.5/4.5 = 1$, the third row is your pivot row. You then interchange it with the second row. This produces a new array

$$A = \begin{bmatrix} -2 & 1 & 3 & -3 \\ 0 & 4.5 & .5 & 8.5 \\ 0 & 2.5 & 11.5 & -6.5 \end{bmatrix}.$$

Notice that row 1 was left unchanged. Now apply Gaussian elimination to the second and third row (multiply the second row by $-2.5/4.5$ and add to the third row) which yields

$$A = \begin{bmatrix} -2 & 1 & 3 & -3 \\ 0 & 4.5 & .5 & 8.5 \\ 0 & 0 & 11.22 & -11.22 \end{bmatrix}.$$

You are now finished with employing the pivot method (in this particular case), since you can only do it $n-1$ times; in this case $n-1 = 3-1 = 2$. It will follow that the last row of A will give the value of x_n. Here you have the equation

$$11.22x_3 = -11.22$$

so $x_3 = -1$. You now need to divide the third row by $A(3,3)$, giving

$$A = \begin{bmatrix} -2 & 1 & 3 & -3 \\ 0 & 4.5 & .5 & 8.5 \\ 0 & 0 & 1 & -1 \end{bmatrix}.$$

This constitutes the first step in backtracking through the equations to arrive at the known solution to the system. After dividing through the third row by $A(3,3)$, you multiply the third row by

$-A(2,3)$, add to the second row (this forces $A(2,3) = 0$), and then divide the second row by $A(2,2)$, giving

$$A = \begin{bmatrix} -2 & 1 & 3 & -3 \\ 0 & 1 & 0 & 2 \\ 0 & 0 & 1 & -1 \end{bmatrix}.$$

Similarly, you can multiply the third row by $-A(1,3)$ and add to row 1, and then multiply row 2 by $-A(1,2)$ and add to row 1. Then divide row 1 by $A(1,1)$, finally giving the array

$$A = \begin{bmatrix} 1 & 0 & 0 & 1 \\ 0 & 1 & 0 & 2 \\ 0 & 0 & 1 & -1 \end{bmatrix}.$$

The solutions to the initial system are located in the right-most column of A,

$$\begin{bmatrix} 1 \\ 2 \\ -1 \end{bmatrix}$$

so $x_1 = A(1,n+1) = 1$, $x_2 = A(2,n+1) = 2$ and $x_3 = -1$.

This entire process can be broken down into two giant steps. The first giant step (call it PIVOT) is to take the initial array A, which is of the form

$$A = \begin{bmatrix} A(1,1) & A(1,2) & \cdots & A(1,n) & A(1,n+1) \\ A(2,1) & A(2,2) & \cdots & A(2,n) & A(2,n+1) \\ \bullet & & & & \\ \bullet & & & & \\ \bullet & & & & \\ A(n,1) & A(n,2) & \cdots & A(n,n) & A(n,n+1) \end{bmatrix}$$

and convert it (by pivoting) to an array of the form

$$A = \begin{bmatrix} x & x & \cdots & x & x \\ 0 & x & \cdots & x & x \\ 0 & 0 & x & \cdots & x \\ 0 & 0 & \cdots & 0 & x \end{bmatrix}$$

where the x's signify real numbers, and where there are zeros in the lower half of the array; i.e., $A(i,j) = 0$ for $j < i$. The second giant step (call it BACKTRACK) is to change this latter array into one of the form

$$A = \begin{bmatrix} 1 & 0 & 0 \cdots 0 & A(1,n+1) \\ 0 & 1 & 0 \cdots 0 & A(2,n+1) \\ 0 & 0 & 1 \cdots 0 & A(3,n+1) \\ 0 & 0 & 0 \cdots 1 & A(n,n+1) \end{bmatrix}.$$

The solutions to the system are the numbers $A(1,n+1)$, $A(2,n+1)$, •••, $A(n,n+1)$. The following program may help to explain BACKTRACK.

Program

```
100   REM BACKTRACK
110   A(N,N+1) = A(N,N+1)/A(N,N)
120   A(N,N) = 1
130   FOR STEP = 1 TO (N−1)
140     ROW = N − STEP
150     FOR COLUMN = (ROW + 1) TO N
160     A(ROW,N+1)=−A(ROW,COLUMN)
          * A(COLUMN,N+1) + A(ROW,N+1)
170     A(ROW,COLUMN) = 0
180     NEXT COLUMN
190     A(ROW,N+1) = A(ROW,N+1)/A(ROW,ROW)
200     A(ROW,ROW) = 1
210   NEXT STEP
220   REM OUTPUT SOLUTIONS
230   FOR SOLUTIONS = 1 TO N
240     PRINT"X(";SOLUTION;")="; A(SOLUTION,N+1)
250   NEXT SOLUTION
```

Exercises

1. Write a program that uses Gaussian elimination with pivoting to solve a system of n linear equations in n unknowns. Test your program with the following two systems.

$$1. -2x_1 + x_2 + 3x_3 = -3 \qquad x_1 + 4x_2 - x_3 = 10$$
$$3x_1 + x_2 + 7x_3 = -2$$

2. $\begin{aligned} 2x_1 - x_2 - x_3 + x_4 + 3x_5 &= 5 \\ -4x_1 + x_2 + x_3 - x_4 + 2x_5 &= -1 \\ 2x_1 + 6x_2 - 3x_3 + 5x_4 - 8x_5 &= 14 \\ x_2 + 5x_3 - x_4 + 4x_5 &= -9 \\ 6x_1 \qquad\quad - 2x_3 + 3x_4 - 5x_5 &= 4. \end{aligned}$

2. Solve the system,

$$x_1 + x_2 = 2$$
$$\beta x_1 + x_2 = 2 + \beta.$$

What is the approximate solution to this system when $\beta = 1 - 10^{-10}$? What kind of computer solution do you think you'll get using this value of β?

SPLINES

Many times an analysis of a real life physical phenomena, such as the data compiled on the properties of air flow around an airplane wing or an automobile hood with a particular shape, or the solution to some partial differentiation equation, reduces to an approximation of the solution with finitely many points. For the remainder of this section suppose that you are dealing with such an unknown function, say f, of a single real variable x; so $y = f(x)$. You will be given a set of n points (x_1,y_1), (x_2,y_2), $\cdots$, (x_n,y_n) that satisfy the function and furnish the only description of f. You can assume that the values of x have been ordered so that $x_1 < x_2 < \cdots < x_n$. The question is, how can you use this data to obtain a meaningful approximation for f?

The objective will be to determine a function g which is satisfied by all the points (x_i,y_i), and is, in some sense, a good approximate to f. See Fig. 4-19.

You know from geometry that any two points in the plane determine a unique linear function; any three points (all with different x coordinates) determine a unique interpolating quadratic function; and any four points determine a unique cubic function. It shouldn't be too surprising to believe that this pattern generalizes: thus, for no points (x_1,y_1), (x_2,y_2), $\cdots$, (x_n,y_n) with $x_i < x_{i+1}$, there exists a unique interpolating polynomial function p_{n-1} of degree $n-1$ for which

$$p_{n-1}(x_i) = y_i$$

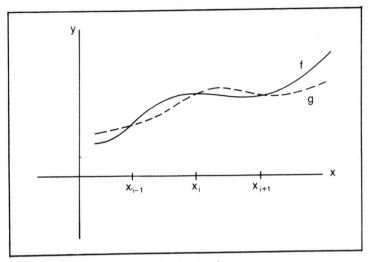

Fig. 4-19. Function g approximating function f.

for all i. This result is known as the Lagrange Interpolation Theorem [63]. It might prove instructive to exhibit the construction process of p_{n-1}. This polynomial function is of the form

$$p_{n-1}(x) = y_1 h_1(x) + y_2 h_2(x) + \cdots + y_n h_n(x)$$

where each function $h_i(x)$ is a polynomial of degree $n-1$ and is of the form

$$h_i(x) = \frac{(x - x_1)(x - x_2)\cdots(x - x_{i-1})(x - x_{i+1})(x - x_{i+2})\cdots(x - x_n)}{(x_i - x_1)(x_i - x_2)\cdots(x_i - x_{i-1})(x_i - x_{i+1})(x_i - x_{i+2})\cdots(x_i - x_n)}.$$

This can be illustrated by forming the polynomial p_3 that passes through the four points $(-1, -2)$, $(0,1)$, $(1, -1)$, $(2,2)$. The four corresponding functions h_i are,

$$h_1(x) = \frac{(x - x_2)(x - x_3)(x - x_4)}{(x_1 - x_2)(x_1 - x_3)(x_1 - x_4)} = \frac{(x - 0)(x - 1)(x - 2)}{(-1 - 0)(-1 - 1)(-1 - 2)}$$

$$= -\frac{1}{6}(x)(x - 1)(x - 2)$$

$$h_2(x) = \frac{(x + 1)(x - 1)(x - 2)}{(0 + 1)(0 - 1)(0 - 2)} = \frac{1}{2}(x + 1)(x - 1)(x - 2)$$

$$h_3(x) = \frac{(x + 1)(x - 0)(x - 2)}{(1 + 1)(1 - 0)(1 - 2)} = \frac{1}{2}(x + 1)(x)(x - 2)$$

$$h_4(x) = \frac{1}{6}(x + 1)(x)(x - 1)$$

and thus

$$p_3(x) = -2h_1(x) + h_2(x) - h_3(x) + 2h_4(x)$$

$$= \frac{1}{6}(10x^3 - 15x^2 - 7x + 6).$$

This function is pictured in Fig. 4-20.

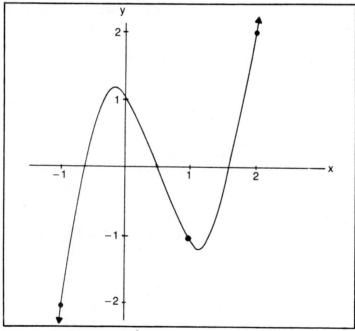

Fig. 4-20. Graph of $p_3(x) = \frac{1}{6}(10x^3 - 15x^2 - 7x + 6)$.

The truly important question is, how close an approximate p_{n-1} is to f? This is difficult to answer, unless more information is known concerning f. We would like to believe that f is at least continuous. But, if you go further and assume much more smoothness, such as assuming f has at least n derivatives (so $f^{(n)}(x)$ exists), some estimates can be made concerning the error $E = ||f - p_{n-1}||$. If this error E is defined by

$$||f - p_{n-1}|| = \max_{x} |f(x) - p_{n-1}(x)|$$

then one estimate [63] states that the error between f and p_{n-1} is given by

$$||f - p_{n-1}|| = \frac{||f^{(n)}|| [\max_{i}(x_{i+1} - x_i)]^n}{4n}$$

From this you learn that it is possible for the error to be large if f is highly oscillatory (e.g., $||f^{(n)}||$ is great). If this is the case, a large number of points (also called knots) for x_i would be needed so that the interpolating polynomial p_{n-1} could furnish a better approximate to f. But, this is exactly the source of the problem! It may be impractical to choose a large n, especially because of the difficulty of working with a polynomial of large degree. Several sections of this book have already been devoted to solving polynomials of degree three, and quite a bit of work is involved. Imagine, instead, the difficulties of working with a polynomial of degree 20 or 50. One means of overcoming these difficulties is to use piecewise fits to f. This leads us into the study of splines!

A spline function is a function consisting of polynomial pieces on subintervals joined together with certain smoothness conditions. One such function is depicted in Fig. 4-21. This continuous function, denoted by f, is defined on [0,5], and consists of five linear functions separately defined on the subintervals [0,1], [1,2], [2,3], [3,4], and [4,5]. You could describe f more accurately by writing

$$f(x) = \begin{cases} -x + 1 & \text{if } x \in [0,1] \\ x - 1 & \text{if } x \in [1,2] \\ 1 & \text{if } x \in [2,3] \\ 2x - 5 & \text{if } x \in [3,4] \\ -x + 7 & \text{if } x \in [4,5] \end{cases}$$

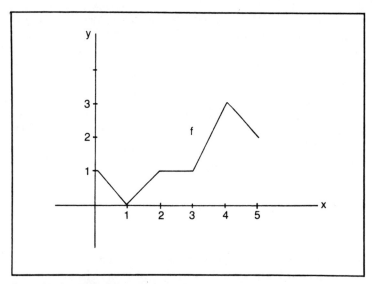

Fig. 4-21. A specific linear spline function.

which defines each of the five connecting linear functions. Since the polynomial pieces are all linear, f is called a linear spline function or a first degree spline function. It is almost always the case that a first-degree spline function is continuous. It is constructed that way. It wouldn't make much sense otherwise.

In general, a linear spline function f is characterized by the following properties:

1. The domain of f is a closed interval [a,b].
2. f is continuous on [a,b].
3. The interval [a,b] is finitely partitioned $a = x_1 < x_2 < \cdots < x_n = b$ (the x_i are called the *knots* of f) so that f, defined on $[x_i,x_{i+1}]$ is a linear function.

The equation for a general first-degree spline function is

$$f(x) = \begin{cases} m_1x + b_1 & \text{if } x \in [x_1,x_2] = [a,x_2] \\ m_2x + b_2 & \text{if } x \in [x_2,x_3] \\ \vdots \\ m_{n-1}x + b_{n-1} & \text{if } x \in [x_{n-1},x_n] = [x_{n-1},b] \end{cases}$$

185

where the m_i's denote the slopes of the individual lines. In order for f to be continuous at all the knots, you will need

$$m_{i-1}x_i + b_{i-1} = m_i x_i + b_i$$

for each i.

An application of linear splines is immediately apparent to any student who has been versed in integral calculus because, if F(x) is any "nice" function defined on [a,b], the trapezoidal rule asserts that the integral of F is approximately equal to the integral of f,

$$\int_a^b F(x)dx \doteq \int_a^b f(x)dx$$

where f is a first-degree spline approximate to F.

As nice as linear splines are, they suffer from a lack of smoothness. This term is used in the mathematical sense to indicate or measure the continuity of a function's derivatives. It is said that an arbitrary function g has smoothness of the order k if the derivative $g^{(k)}(x)$ is continuous and $g^{(k+1)}(x)$ is not continuous. The function $g(x) = |x|$ has smoothness of order 0 because g is continuous while $g'(x)$ fails to exist at $x = 0$. Also, $g(x) = x^{4/3}$ has smoothness of order 1.

Higher degree splines are used whenever you desire more smoothness. Normally cubic splines are preferred. In this case, cubic polynomials are joined together at the knots so that not only do successive polynomials form a continuous curve, but they have equal first and second derivatives at the knots. Because of this, the human eye would be unable to discern any noticeable change in curvature as you pass from one side of the knot x_i to the other side. See Fig. 4-22.

The remainder of this section will be devoted to second-degree splines, which are splines that consist of quadratic polynomials connected together so as to form a continuous curve in which successive quadratics, q_i, have the same slope at the corresponding knot.

To this end, consider a set of n knots,

$$x_1 < x_2 < \bullet\bullet\bullet < x_n$$

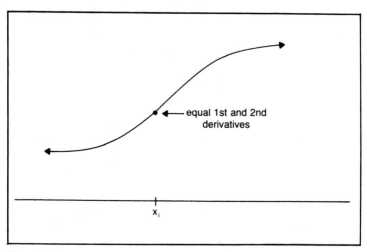

Fig. 4-22. Smoothness of cubic splines at a knot.

not necessarily equally spaced, and a corresponding set of n y-values, y_1, y_2, $\bullet\bullet\bullet$ y_n. The spline function f will be of the form

$$f(x) = \begin{cases} q_1(x) & \text{if } x \in [x_1, x_2] \\ q_2(x) & \text{if } x \in [x_2, x_3] \\ \bullet \\ \bullet \\ \bullet \\ q_{n-1}(x) & \text{if } x \in [x_{n-1}, x_n] \end{cases}$$

where each q_i is a quadratic polynomial, $q_i(x) = a_i x^2 + b_i x + c_i$ as shown in Fig. 4-23.

Initially, you should determine the quadratic q_1 that passes through (x_1, y_1) and (x_2, y_2) and has a given slope m_1 at (x_1, y_1). This slope is arbitrarily selected and fed into your program. Once q_1 is determined, we proceed to find q_2: in this case q_2 is unique because (x_2, y_2) and (x_3, y_3) are known, and so is the slope m_2 of q_2 at (x_2, y_2), since it is equal to $q_1'(x_2)$. The pattern continues with $q_3, q_4, \bullet\bullet\bullet, q_{n-1}$. Each q_i is uniquely determined because it must pass through two given points, and it must possess a known derivative at x_i.

It follows, although I am omitting the details, that the necessary quadratic polynomials, $q_i(x) = a_i x^2 + b_i x + c_i$, are described by the relationships,

187

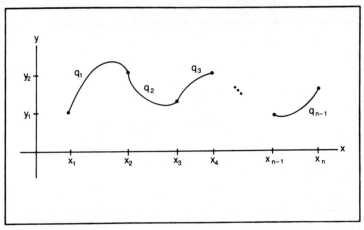

Fig. 4-23. Quadratic spline.

$$1. \ a_i = \frac{m_{i+1} - m_i}{2(x_{i+1} - x_i)}$$

$$2. \ b_i = m_i - \frac{x_i(m_{i+1} - m_i)}{(x_{i+1} - x_i)}$$

$$3. \ c_i = \frac{x_i^2(m_{i+1} - m_i)}{2(x_{i+1} - x_i)} - m_i x_i + y_i$$

where, once m_1 is selected, the remaining slopes can be found by the recursive definition,

$$m_{i+1} = \frac{2(y_{i+1} - y_i)}{(x_{i+1} - x_i)} - m_i \quad \text{for } i = 1,2,\cdots,n-1.$$

Let's work through a particular example and see how these relationships function. Choose six points ($n = 6$) given by $(-1,2)$, $(0,1)$, $(1/2,0)$, $(1,1)$, $(2,2)$, and $(5/2,3)$ and select $m_1 = 1$. Then the remaining slopes are,

$$m_2 = \frac{2(-1)}{1} - 1 = -3$$

$$m_3 = \frac{2(-1)}{1/2} + 3 = -1$$

$$m_4 = 5$$

$$m_5 = -3$$

$$m_6 = 7.$$

This information, so far, means that the $6-1 = 5$ quadratic polynomials must fit into the diagram sketched in Fig. 4-24. The remaining unknowns a_i, b_i, and c_i can be computed from (1),(2), and (3) and are listed in Table 4-9.

The quadratic spline f that fits these 6 data points and has an initial slope $m_1 = 1$ at $x_1 = -1$ is therefore given by the equation,

$$f(x) = \begin{cases} -2x^2 - 3x + 1 & \text{if } x \in [-1,0] \\ 2x^2 - 3x + 1 & \text{if } x \in [0,1/2] \\ 6x^2 - 7x + 2 & \text{if } x \in [1/2,1] \\ -4x^2 - 3x - 8 & \text{if } x \in [1,2] \\ 10x^2 - 43x + 48 & \text{if } x \in [2,5/2]. \end{cases}$$

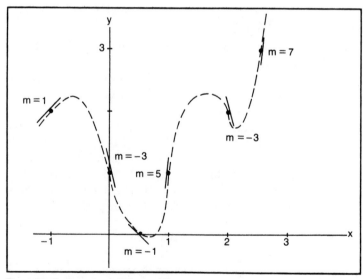

Fig. 4-24. Outline of a quadratic spline.

Table 4-9. Coefficients for a Quadratic Spline.

i	a_i	b_i	c_i
1	-2	-3	1
2	2	-3	1
3	6	-7	2
4	-4	-3	-8
5	10	-43	48

You should sketch this curve and compare it with the outline pictured in Fig. 4-24.

Exercises

1. Write a program that computes the equation of a second-degree spline function f when the points (x_1,y_1), $\cdots$, (x_n,y_n) are fed into the program (make sure $x_i < x_{i+1}$), and when m_1 is also given. Test your program for the situation where $n = 10$, $(x_i,y_i) = (i, \sin(i))$, and $m_1 = 1$.

2. Find the Lagrange polynomial of degree 4 that interpolates the points $(-1,1)$, $(0,2)$, $(1,1)$, $(2,-2)$ and $(3,5)$.

MATRIX SQUARE ROOTS

One of the advantages that many versions of BASIC have over FORTRAN is the special adaptions that allow for easier matrix manipulations. The MAT INPUT, MAT PRINT, and MAT READ statements all prove very useful, as do the MAT functions of TRN, INV, and DET. It is, however, an educational experience to write these programs and not to make use of the special functions. It is the purpose of this section to discuss a means for computing the square root of certain matrices without using any of these BASIC functions.

First the standard method for extracting the square root of an n × n square matrix A will be discussed. The n solutions (real or complex) to the polynomial (the characteristic polynomial)

$$p(\lambda) = \det(A - \lambda I)$$

of degree n, where λ is the independent variable and I is the n×n identity matrix, are to be denoted by $\lambda_1, \lambda_2, \cdots, \lambda_n$. These numbers are known as the *eigenvalues* of the matrix A. Corresponding to each eigenvalue λ_i is a vector v_i (an $n \times 1$ array) that satisfies the equation $Av_i = \lambda_i v_i$. If you write this vector v_i as

$$v_i = \begin{pmatrix} x_{1i} \\ x_{2i} \\ \cdot \\ \cdot \\ \cdot \\ x_{ni} \end{pmatrix}$$

then the above equation becomes

$$\begin{pmatrix} a_{11} & a_{12} & \cdots & a_{1n} \\ a_{21} & a_{22} & \cdots & a_{2n} \\ & & & \\ a_{n1} & a_{n2} & \cdots & a_{nn} \end{pmatrix} \begin{pmatrix} x_{1i} \\ x_{2i} \\ \\ x_{ni} \end{pmatrix} = \lambda_i \begin{pmatrix} x_{1i} \\ x_{2i} \\ \\ x_{ni} \end{pmatrix}$$

You can then form the $n \times n$ matrix P in which the columns of P are the corresponding vectors v_i, which are known as the eigenvectors. If P is invertible, which means it is nonsingular, and must therefore have a nonzero determinant, then P^{-1} exists, and the product,

$$P^{-1} \cdot A \cdot P$$

is a matrix which is diagonal. Furthermore, the elements along the diagonal are precisely the eigenvalues λ_i of A. If you set D equal to this diagonal matrix, then the square root of D, denoted by $D^{1/2}$, is simply that matrix (there are 2^n possibilities) with $\pm \sqrt{\lambda_i}$ along the main diagonal. For example, if D is the 2×2 matrix

$$D = \begin{pmatrix} 4 & 0 \\ 0 & 9 \end{pmatrix}$$

then there are four square roots of D, given by

$$1. \begin{pmatrix} 2 & 0 \\ 0 & 3 \end{pmatrix} \quad 2. \begin{pmatrix} 2 & 0 \\ 0 & -3 \end{pmatrix}$$

$$3. \begin{pmatrix} -2 & 0 \\ 0 & 3 \end{pmatrix} \quad 4. \begin{pmatrix} -2 & 0 \\ 0 & -3 \end{pmatrix}.$$

Finally, if you set C equal to the product, $C = P \cdot D^{1/2} \cdot P^{-1}$, for any choice of $D^{1/2}$, C is a square root of A. Keep in mind that what this means is $C^2 = A$.

Trace through some sample calculations by first setting A equal to the 3×3 matrix

$$A = \begin{pmatrix} 4 & 1 & 0 \\ 1 & 4 & 0 \\ 0 & 0 & 2 \end{pmatrix}.$$

The characteristic polynomial $p(\lambda)$ is the determinant of the array

$$\begin{pmatrix} 4-\lambda & 1 & 0 \\ 1 & 4-\lambda & 0 \\ 0 & 0 & 2-\lambda \end{pmatrix}$$

which gives $p(\lambda) = (\lambda^2 - 8\lambda + 15)(2 - \lambda) = (2 - \lambda)(3 - \lambda)(5 - \lambda)$. The eigenvalues are thus $\lambda_1 = 2$, $\lambda_2 = 3$ and $\lambda_3 = 5$. The corresponding eigenvectors v_i are

$$v_1 = \begin{pmatrix} 0 \\ 0 \\ 1 \end{pmatrix}, \quad v_2 = \begin{pmatrix} 1 \\ -1 \\ 0 \end{pmatrix}, \quad v_3 = \begin{pmatrix} 1 \\ 1 \\ 0 \end{pmatrix},$$

which means the matrix P is given by

$$P = \begin{pmatrix} 0 & 1 & 1 \\ 0 & -1 & 1 \\ 1 & 0 & 0 \end{pmatrix}.$$

This matrix is invertible (its determinant is one) and the inverse P^{-1} is

$$P^{-1} = \begin{pmatrix} 0 & 0 & 1 \\ 1/2 & -1/2 & 0 \\ 1/2 & 1/2 & 0 \end{pmatrix}.$$

It follows that $D = P^{-1} \cdot A \cdot P$ is equal to

$$D = \begin{pmatrix} 2 & 0 & 0 \\ 0 & 3 & 0 \\ 0 & 0 & 5 \end{pmatrix}.$$

There are eight square roots, $D^{1/2}$, of D, and if you select the one with all positive diagonal elements, then $C = P \cdot D^{1/2} \cdot P^{-1}$ has the value

$$C = \begin{pmatrix} 0 & 1 & 1 \\ 0 & -1 & 1 \\ 1 & 0 & 0 \end{pmatrix} \begin{pmatrix} \sqrt{2} & 0 & 0 \\ 0 & \sqrt{3} & 0 \\ 0 & 0 & \sqrt{5} \end{pmatrix} \begin{pmatrix} 0 & 0 & 1 \\ 1/2 & -1/2 & 0 \\ 1/2 & 1/2 & 0 \end{pmatrix}$$

$$= \frac{1}{2} \begin{pmatrix} \sqrt{3} + \sqrt{5} & -\sqrt{3} + \sqrt{5} & 0 \\ -\sqrt{3} + \sqrt{5} & \sqrt{3} + \sqrt{5} & 0 \\ 0 & 0 & 2\sqrt{2} \end{pmatrix}.$$

Thus, you have obtained a square root of A.

To simplify matters, let's consider only those matrices A of order 2×2. This restriction would certainly make a lot of the above computations easier, particularly if being done by hand (it might not make much difference to the computer). Of course it is a lot easier to solve a characteristic polynomial of degree two than degree three! An interesting reference [48] that provides a fresh approach to obtaining square roots of 2×2 arrays appeared last year.

A summary of this article concludes with the relationship

$$C = \frac{1}{\text{Tr}(C)} \left[A \pm (\det A)^{1/2} \cdot I \right]$$

where Tr(C), the trace of C, is the sum of the diagonal elements of C. Furthermore, it follows that the trace of C is

$$\text{Tr}(C) = \pm (\text{Tr}(A) \pm 2(\det A)^{1/2}).$$

To simplify matters, suppose A is the array given by

$$A = \begin{pmatrix} a & b \\ c & d \end{pmatrix}$$

so $\mathrm{Tr}(A) = a + d$ and $\det A = ad - bc$. If you set $\theta = |(\det A)^{1/2}|$, the above relationships reduce to the following four square roots of A:

1. $C_1 = \dfrac{1}{\sqrt{a + d + 2\theta}} \begin{pmatrix} a+\theta & b \\ c & d+\theta \end{pmatrix}$

2. $C_2 = \dfrac{1}{\sqrt{a + d - 2\theta}} \begin{pmatrix} a-\theta & b \\ c & d-\theta \end{pmatrix}$

3. $C_3 = \dfrac{-1}{\sqrt{a + d + 2\theta}} \begin{pmatrix} a+\theta & b \\ c & d+\theta \end{pmatrix}$

4. $C_4 = \dfrac{-1}{\sqrt{a + d - 2\theta}} \begin{pmatrix} a-\theta & b \\ c & d-\theta \end{pmatrix}$

This reduction makes it easy for you to program the computation of $A^{1/2}$ unless $\det A < 0$ or $a + d \pm 2\theta \leq 0$, and then problems occur!

Before we enter a discussion on what to do when one of these trouble spots pop-up, let us work through an example in which all the numbers involved are real numbers. If you set A equal to the array

$$A = \begin{pmatrix} 5 & 2 \\ 2 & 4 \end{pmatrix}$$

then $\det A = 16$, $\mathrm{Tr}(A) = 9$, and $\theta = 4$. The four square roots of A are:

1. $C_1 = \dfrac{1}{\sqrt{17}} \begin{pmatrix} 9 & 2 \\ 2 & 8 \end{pmatrix}$

2. $C_2 = \dfrac{1}{1} \begin{pmatrix} 1 & 2 \\ 2 & 0 \end{pmatrix}$

3. $C_3 = \dfrac{-1}{\sqrt{17}} \begin{pmatrix} 9 & 2 \\ 2 & 8 \end{pmatrix}$

$$4. \quad C_4 = \frac{-1}{1} \begin{pmatrix} 1 & 2 \\ 2 & 0 \end{pmatrix}.$$

You are not always going to have four square roots of A; you may have only two square roots, no square roots, or infinitely many square roots. The real problem arises when there is division by zero, when $a + d + 2\theta = 0$, or when $a + d - 2\theta = 0$. This section will examine these three cases, assuming in all three cases that $|a + d| = 2|\theta|$.

Case 1. $a + d = 0$ and $\theta = 0$.

This situation can occur in many different ways, for example,

$$A_1 = \begin{pmatrix} 4 & -4 \\ 4 & -4 \end{pmatrix}, \quad A_2 = \begin{pmatrix} 0 & 0 \\ 3 & 0 \end{pmatrix}, \quad A_3 = \begin{pmatrix} 0 & 0 \\ 0 & 0 \end{pmatrix}$$

but as long as A contains at least one nonzero element, A will not have a square root. Only the zero matrix, A_3, has a square root—namely itself.

Case 2. $a + d \neq 0$, $\theta \neq 0$, $b \neq 0$, $c \neq 0$.

Either $a + d = 2\theta$ or $a + d = -2\theta$; suppose the former. Since C_2 and C_4 fail to exist (because $a + d - 2\theta = 0$), only C_1 and C_3 represent square roots.

Case 3. $a + d \neq 0$, $\theta \neq 0$, $b = 0$, $c = 0$.

Since $|a + d| = 2|\theta| = 2\sqrt{ad}$, squaring both sides and simplifying gives $(a - d)^2 = 0$, or $a = d$. Thus

$$A = \begin{pmatrix} a & b \\ c & d \end{pmatrix} = \begin{pmatrix} a & 0 \\ 0 & a \end{pmatrix} = a \cdot I$$

and consequently

$$C = a^{1/2} \cdot I^{1/2}.$$

If $a < 0$, C will be imaginary; otherwise C is real. But what about $I^{1/2}$, the square root of the identity matrix? Do you know what $I^{1/2}$ is? Oddly enough, there are an infinite number of square roots of I. Some of these include,

195

$$\begin{pmatrix} 1 & 2 \\ 0 & -1 \end{pmatrix}, \begin{pmatrix} 1 & 1 \\ 0 & -1 \end{pmatrix}, \begin{pmatrix} 2 & 1 \\ -3 & -2 \end{pmatrix}, \begin{pmatrix} 1 & 0 \\ 0 & 1 \end{pmatrix}$$

and in fact, any array of the form

$$\begin{pmatrix} e & f \\ \dfrac{1-e^2}{f} & -e \end{pmatrix}$$

is a square root of I.

For programming brevity, terminate computations as soon as you know C is to involve imaginary numbers. This happens, first, if det A $<$ 0, and second, if det A $>$ 0 but a + d $\pm$ 2θ $<$ 0. You only want to print out real square roots. For given values of a, b, c, and d the computation of $A^{1/2}$ is not involved, although it does involve several decision statements, as the flowchart in Fig. 4-25 depicts.

Exercises

1. Write a program that computes the real square roots of a given matrix A of the form,

$$A = \begin{pmatrix} a & b \\ c & d \end{pmatrix}.$$

If A has two or four square roots, print them all out. If A has infinitely many square roots, print out this fact and exhibit two of them. Test your program on the following samples.

1. $A = \begin{pmatrix} 5 & 1 \\ -1 & 3 \end{pmatrix}$ 3. $A = \begin{pmatrix} 2 & -1 \\ 4 & -2 \end{pmatrix}$

2. $A = \begin{pmatrix} 5 & 6 \\ -1 & 2 \end{pmatrix}$ 4. $A = \begin{pmatrix} 3 & 0 \\ 0 & 3 \end{pmatrix}$

2. Suppose the matrix A, given by

$$A = \begin{pmatrix} 1 & b \\ 2 & d \end{pmatrix}$$

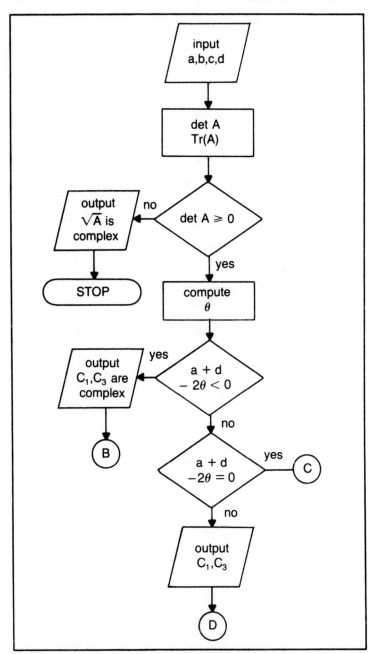

Fig. 4-25. Flowchart for computing square root of 2 × 2 matrix.

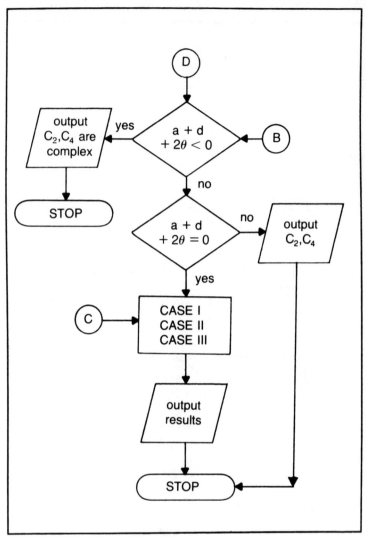

Fig. 4-25. Continued from page 197.

has two eigenvalues, say λ_1 and λ_2. Find b and d.

THE GRAM-SCHMIDT PROCESS

Suppose you have a given vector space, E, and a real inner product function f defined on E×E. Recall that this means that for any two elements u, v from E, f(u,v) is a uniquely determined real number for which f satisfies the properties,

1. $f(u,v) = f(v,u)$ (symmetry)
2. $f(\lambda_1 u + \lambda_2 v, w) = \lambda_1 f(u,w) + \lambda_2 f(v,w)$ (linearity)
3. $f(u,u) \geq 0$
4. $f(u,u) = 0$ if and only if $u = 0$.

Most texts customarily denote f by the symbol $<,>$, thus giving us the notation

$$f(u,v) = <u,v>.$$

Several standard examples of vector spaces equipped with an inner product include the following.

Example 1. Set E equal to 2-space, $E = R^2$, where a typical element $u \in E$ is a vector represented by the two coordinates u_1 and u_2. Therefore $u = (u_1, u_2)$, where each coordinate u_i is a real number. Then $f(u,v)$ is defined by

$$f(u,v) = u_1 v_1 + u_2 v_2.$$

Example 2. Set E equal to 4-space, $E = R^4$, where $u \in E$ is a vector given by $u = (u_1, u_2, u_3, u_4)$. Then $f(u,v) = <u,v>$ is given by

$$<u,v> = u_1 v_1 + 2u_2 v_2 + 3u_3 v_3 + 4u_4 v_4.$$

Example 3. Set E equal to $P_n [0,1]$, the space of polynomials of degree at most n defined on the unit interval. Then if $p, q \in E$ we define the inner product by

$$<p,q> = \int_0^1 p(x)q(x)dx.$$

Note that this last integral definition satisfies the necessary conditions for an inner product because

1. $\int_0^1 p(x)q(x)dx = \int_0^1 q(x)p(x)dx$

2. $\int_0^1 (\lambda_1 p(x) + \lambda_2 q(x))r(x)dx = \lambda_1 \int_0^1 p(x)r(x)dx + \lambda_2 \int_0^1 q(x)r(x)dx$

3. $\int_0^1 p(x)p(x)dx \geq 0$

4. $\int_0^1 p^2(x)dx = 0$ if and only if $p = 0$.

You can verify that the inner product conditions are satisfied for the first two examples.

One very interesting concept that develops from an inner product is the notion of the norm $||u||$ of an arbitrary vector u ε E. This definition is given by

$$||u|| = <u,u>^{1/2}.$$

In the case $E = R^2$ with the aforementioned standard inner product if $u = (2,3)$, then $||u|| = (4+9)^{1/2} = \sqrt{13}$. If $E = P_2[0,1]$ with $p(x) = 1 + x + x^2$, then the norm of p is

$$||p|| = \left[\int_0^1 (1 + x + x^2)^2 dx \right]^{1/2}$$

$$= \left[\int_0^1 (1 + 2x + 3x^2 + 2x^2 + x^4)dx \right]^{1/2}$$

$$= \sqrt{3.7}.$$

The importance of the norm is that it furnishes a generalization of the concept of distance. Whereas in ordinary Euclidean geometry, the distance between two points $u = (u_1,u_2)$, $v = (v_1,v_2)$ is given by

$$||u - v|| = [(u_1 - v_1)^2 + (u_2 - v_2)^2]^{1/2},$$

in an inner product space, the distance between two points u, v is given by the norm of their difference, $||u - v||$. Thus, in the three spaces given above, you have

1. $||u - v|| = <u-v, u-v>^{1/2}$
    $= [(u_1 - v_1)^2 + (u_2 - v_2)^2]^{1/2}$

2. $| |u - v| | = <u-v, u-v>^{1/2}$
$$= [(u_1 - v_1)^2 + 2(u_2 - v_2)^2 + 3(u_3 - v_3)^2 + 4(u_4 - v_4)^2]^{1/2}$$
3. $| |p - q| | = <p-q, p-q>^{1/2}$
$$= \int_0^1 (p(x) - q(x))^2 dx]^{1/2}.$$

You can see that the generalized notion of distance in R^2 agrees with the Euclidean distance. Furthermore, the specific distance in $P_2[0,1]$ between polynomials $p(x) = x$ and $q(x) = x^2$ is

$$| |p - q| | = \left[\int_0^1 (x - x^2)^2 dx \right]^{1/2}$$

$$= \left[\int_0^1 (x^2 - 2x^3 + x^4)dx \right]^{1/2}$$

$$= \sqrt{1/30}.$$

The inner product also serves as a means for determining the angle θ between the two given vectors u, v ϵ E. Here, the standard dot product of two vectors, $u \cdot v = |u| \cdot |v| \cos(\theta)$, is generalized to give

$$<u, v> = | |u| | \cdot | |v| | \cos(\theta)$$

and, solving for θ,

$$\theta = \cos^{-1} \left[\frac{<u, v>}{| |u| | \cdot | |v| |} \right], u \neq 0, v \neq 0.$$

Thus, for vectors, u, v ϵ E $= R^4$ with u $= (1,0,1,-1)$, and v $= (0,2,1,2)$, the angle θ between them is found by

$$\cos(\theta) = \frac{<u, v>}{| |u| | \, | |v| |}$$

$$= \frac{0 + 0 + 3(1) + 4(-2)}{\sqrt{1 + 3 + 4} \, \sqrt{8 + 3 + 16}}$$

$$= \frac{-5}{6\sqrt{6}}$$

so

$$\theta = \cos^{-1}(-5\sqrt{6}/36) = 1.917 \text{ rad} = 109.9°.$$

Because $|\cos(\theta)| \leq 1$ for all θ, it follows in general that

$$|<u, v>| \leq |\ |u|\ |\bullet|\ |v|\ |$$

which is the celebrated Cauchy-Schwarz inequality.

In Euclidean 2-space or 3-space, vectors are said to be perpendicular (orthogonal) if the angle θ formed between them is 90°; this means $\cos(\theta) = 0$. Correspondingly, in an inner product space, two vectors, u, v are said to be orthogonal, as shown in Fig. 4-26, if $<u, v> = 0$. It follows that in $P_2[0,1]$, the polynomials $p(x) = x^2 - 1$ and $q(x) = 1 - 5x^2$ are orthogonal.

One last bit of background information will be discussed before the main topic of this section, the Gram-Schmidt process [54], can be addressed. Recall that in a vector space E, a set of elements $S = \{e_1, e_2, \bullet\bullet\bullet, e_n\}$ is a basis for E if the elements e_i are all linearly independent, and the e_i span E. Thus, for any $u \in E$, there exists unique scalars $\lambda_1, \lambda_2, \bullet\bullet\bullet, \lambda_n$ such that $u = \lambda_1 e_1, + \lambda_2 e_2 + \bullet\bullet\bullet + \lambda_n e_n$. In the case $E = R^2$, the two elements $e_1 = (1.0)$ and $e_2 = (0,1)$ form a basis for E. Not only are these two elements independent, but any $u = (u_1, u_2)$ can be written uniquely as $u = u_1 e_1 + u_2 e_2$. The two elements $e_1 = (1,1)$ and e_2 $(0,1)$ could also be used as a basis for R^2.

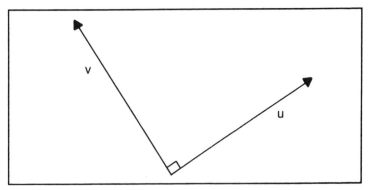

Fig. 4-26. Orthogonal vectors.

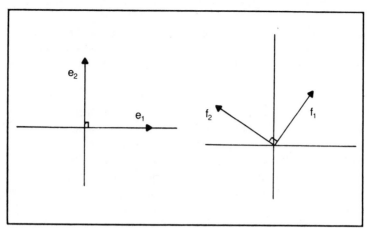

Fig. 4-27. Two orthonormal bases in R^2.

And, if $E = P_2[0,1]$, the elements $e_1 = 1$, $e_2 = x$, and $e_3 = x^2$ form a basis S for E.

Quite often a space E will have many different bases. In $E = R^2$, for example, any two elements $e = (e_1, e_2)$, $f = (f_1, f_2)$ form a basis as long as $e_1 f_2 \neq e_2 f_1$. But some bases have properties more desirable than others! What are these properties? There are two that are especially significant: first, it is desirable that the basis elements e_1, e_2, $\bullet\bullet\bullet$, e_n be mutually orthogonal to one another; and second, it is desirable that each e_i be of unit length. In terms of inner products, these two properties translate into

1. $<e_i, e_j> = 0$ if $i \neq j$
2. $||e_i|| = 1$ for all i.

If E is an inner product space, and S is a basis for E with both of these properties, then S is called an *orthonormal basis*. Now, simply stated, the Gram-Schmidt process is a process which is applied to any basis S to produce an orthonormal basis. This process is named for the Danish actuary Jörgen Gram (1850 - 1916) and the German mathematician Erhardt Schmidt (1876 - 1959).

In the plane, $E = R^2$, the vectors $e_1 = (1,0)$, and $e_2 = (0,1)$ form an orthonormal basis; while the vectors $f_1 = (.5, \sqrt{3}/2)$, and $f_2 = (-\sqrt{3}/2, .5)$ form another orthonormal basis as shown in Fig. 4-27).

The basis $\{1, x, x^2\}$ for $P_2[0,1]$ is not an orthonormal basis because no two of these vectors are orthogonal, and only one of the vectors has unit length. On the other hand, the basis $\{1, 2x, 3x^2\}$

203

contains all unit vectors; unfortunately no two are orthogonal. This leads to the question: how can you construct 3 vectors, all of unit length, that are pairwise orthogonal? The Gram-Schmidt process will show you how.

First consider an inner product space E, and two vectors e_1, and e_2 from an arbitrary basis S. If you let A be the terminal point of e_2, then e_2 is represented by $\overrightarrow{OA}$. Then drop a perpendicular form A down to e_1, as shown in Fig. 4-28. This determines point B. The vector $f_1 = \overrightarrow{OB}$ is called the projection of e_2 into e_1. The vector $f_2 = \overrightarrow{BA}$ is perpendicular to f_1, and hence perpendicular to e_1 (so $<e_1, f_2> = 0$), and is called the projection of e_2 orthogonal to e_1. Since f_1 is some scalar multiple of e_1 ($f_1 = \lambda e_1$), and since $f_1 + f_2 = e_2$, it follows that

$$
\begin{aligned}
<e_1, e_2> &= <e_1, f_1 + f_2> \\
&= <e_1, \lambda e_1 + f_2> \\
&= \lambda <e_1, e_1> + <e_1, f_2>
\end{aligned}
$$

where this last step makes use of the linear property of an inner product. Recalling that e_1 and f_2 are orthogonal, you can solve for λ and get $\lambda = <e_1, e_2>/<e_1, e_2>$. This means

$$
\begin{aligned}
f_1 &= \lambda e_1 \\
&= \frac{<e_1, e_2>}{<e_1, e_1>} e_1
\end{aligned}
$$

and thus

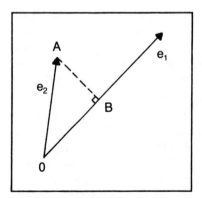

Fig. 4-28. Projection of e_2 onto e_1.

$$f_2 = e_2 - f_1$$

$$= e_2 - \frac{<e_1, e_2>}{<e_1, e_1>} e_1.$$

Now, where is the orthonormal basis? This basis, call it S′, will contain $\{e_1', e_2', \cdots, e_n'\}$ where, for the time being, you know the first two vectors e_1' and e_2'. These are,

1. $e_1' = \dfrac{e_1}{||e_1||}$

2. $e_2' = \dfrac{f_2}{||f_2||} = \dfrac{e_2 - <e_1', e_2> e_1'}{||e_2 - <e_1', e_2> e_1'||}$.

For instance, if $E = R^2$, $e_1 = (1,1)$, and $e_2 = (-1, 2)$; then

$$e_1' = \frac{\sqrt{2}}{2} (1,1)$$

and

$$e_2' = \frac{(-3/2, 3/2)}{3\sqrt{2}/2} = \frac{\sqrt{2}}{3} (-3/2, 3/2).$$

The next vector e_3' in S′ has the form

$$e_3' = \frac{f_3}{||f_3||}$$

where $f_3 = e_3 - <e_1', e_3> e_1' - <e_2', e_3> e_2'$. In general, if $S = \{e_1, e_2, \cdots, e_n\}$ then $S' = \{e_1', e_2', \cdots, e_n'\}$ where $e_i' = f_i/||f_i||$, and

$$f_i = e_i - <e_1', e_i> e_1' - <e_2', e_i> e_2' - \cdots - <e_{i-1}', e_i> e_{i-1}'.$$

This iterative scheme is the Gram-Schmidt process.

Program this process for the particular inner product space $E = R^n$ (for some $n \leq 10$), where, for vectors $u = (u_1, u_2, \cdots, u_n)$, and $v = (v_1, v_2, \cdots, v_n)$, the inner product $<u, v>$ is defined by

$$<u, v> = u_1 v_1 + u_2 v_2 + \cdots + u_n v_n.$$

205

You should note that this definition is a special case of

$$<u, v> = k_1 u_1 v_1 + k_2 u_2 v_2 + \cdots + k_n u_n v_n,$$

namely, when $k_i = 1$ for all i. Basis vectors $e_1, e_2, \cdots, e_n$ will be fed into the program, and so will the value of n. The basis vectors could easily be stored in arrays, say $E(1,n)$, $E(2,n)$, $\cdots$, $E(n,n)$. You will also need to store the vectors $e_1', e_2', \cdots, e_n'$. Computation of each e_i' comes next, and you will need to compute some inner products (in fact, it will require $i-1$ inner products to evaluate e_i'). This might be an appropriate place in your program to build a subroutine or subfunction to evaluate an inner product.

Exercises

1. Write a program that computes this orthonormal basis S', when you are given S, by using the Gram-Schmidt process. Assume the inner product $<u, v>$ is defined by $u_1 v_1 + u_2 v_2 + \cdots u_n v_n$. Test your program with the following bases.

1. $n = 3$	2. $n = 5$
$e_1 = (1,2,1)$	$e_1 = (1,2,1,1,0)$
$e_2 = (-1,0,1)$	$e_2 = (1,0,2,3,-1)$
$e_3 = (1,1,-1)$	$e_3 = (4,1,1,3,-1)$
	$e_4 = (2,2,3,-1,2)$
	$e_5 = (1,0,0,4,1)$

After computing $S' = \{e_1', e_2', \cdots, e_n'\}$, check to make sure each of these vectors has unit length ($<e_i', e_i> = 1$ for each i), and make sure that the vectors are pairwise orthogonal ($<e_i', e_j'> = 0$ if $i \neq j$). Due to roundoff error, it may be that $<e_i', e_j'>$ is approximately but not exactly zero. What kind of tolerance would you be willing to settle for?

2. Consider the inner product space $P_2[0,1]$, and choose $S = \{1,x,x^2\}$. Use the Gram-Schmidt process to compute S'.

Chapter 5
Analysis

The study of analysis is, briefly, the study of function theory. The student typically begins with functions of a single independent variable, f(x), and then progresses to functions of two variables f(x,y), functions of n independent variables, $f(x_1, x_2, \cdots, x_n)$, and finally to functions of infinitely many variables. The domain of these variables usually includes some continuum of values, such as an interval, [a,b], or the entire real line $(-\infty, \infty)$, although it's not uncommon to have a discrete domain.

Why do people study functions in the first place? I'm sure with a little thought you will realize that functions provide a means for expressing how some unknown quantity (a company's monthly profit; the volume of a solid figure; the height of a projectile) varies with respect to other quantities (number of employees, length, width, and height; initial velocity and time). One can compute what the effects will be on the dependent variable when the independent variables are changed by a particular amount.

Certain characteristics of functions are more important than others, and hence these are the ones that scientists are most concerned with. The concepts of continuity and differentiability are two of these most vital properties. Functions that are continuous and differentiable (this is actually repetitious, because if f is differentiable it must be continuous) are considered "nice" functions. The function becomes "nicer" and "smoother" with each successive derivative that it possesses. The function $f(x) = x^{1/3}$ is an example of a continuous function which is differentiable everywhere except at x = 0, while $f(x) = x^{4/3}$ is continuous and differentiable everywhere, but fails to have a second derivative at x = 0. This function is therefore smoother than $f(x) = x^{1/3}$. The functions $y = \sin(x)$, $y = e^x$, and the polynomials are examples of functions with infinitely many derivatives.

The point to be made here is that a function can be classified into one of three groups: those that are nice, those that are somewhat nice, and those that are badly behaved. This phraseology is, of course, very vague, but it's done on purpose to get across a concept. You could define these groups more precisely by, say, focusing in on the differentiability of the function. For example, f would be nice if it were everywhere differentiable; f would be badly behaved if it had uncountably many points lacking differentiability.

I wish to present some examples of functions that fall into this latter class, functions that are so badly behaved they could easily be labeled bizarre. These examples serve a dual purpose: not only do these functions illustrate the existence of functions with such wild properties, but they signal caution to scientists about jumping to hasty conclusions.

The first such function, sometimes known as the salt and pepper function, is simply defined by

$$f(x) = \begin{cases} 1 & \text{if x is rational} \\ 0 & \text{if x is irrational.} \end{cases}$$

The graph of f in Fig. 5-1 consists of two horizontal rows of points. Not only is f without any points of differentiability, it is also nowhere continuous! For those that are more familiar with the terminology of mathematics, this function is nowhere monotonic, not of bounded variation, not absolutely continuous, and not Riemann integrable (but it is Lebesgue integrable).

The second function, known as the ruler function, is defined on the interval [0,1] by

$$f(x) = \begin{cases} 1/q & \text{if } x = p/q, \ (p,q) = 1 \\ 0 & \text{if x is irrational.} \end{cases}$$

Whenever x is a rational number, it is written as the ratio of two integers p/q, where p and q are relatively prime and then f(x) is

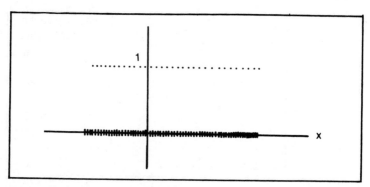

Fig. 5-1. The salt and pepper function.

defined to be the reciprocal of q. The graph of f is pictured in Fig. 5-2. This function happens to be discontinuous at all rational points, hence it is not differentiable there. It is continuous at all irrationals and is even differentiable there.

The third function, the Cantor function on [0,1], named after Georg Cantor, serves mathematicians with a handy counterexample to many seemingly plausible conjectures. First, any number $x \in [0,1]$ can be expressed in decimal form in a base-3 expansion. Thus, $x = .x_1x_2x_3\cdots$ where $x_1 \in \{0,1,2\}$. The Cantor function f is then defined by

$$f(x) = \begin{cases} .\dfrac{x_1}{2}\dfrac{x_2}{2}\dfrac{x_3}{2}\cdots & \text{(base 2)} \quad \text{if } x_i \in \{0,2\} \text{ for all } i \\[4mm] .\dfrac{x_1}{2}\dfrac{x_2}{2}\cdots\dfrac{x_n}{2} & \text{(base 2)} \quad \text{if } x_i \in \{0,2\} \text{ for } i = 1,2,\cdots,n-1, \\ & \qquad\qquad\qquad \text{and } x_n = 1. \end{cases}$$

The graph of f in Fig. 5-3 has the appearance of an infinite set of stairsteps. The Cantor function is everywhere continuous and is monotone (which is a protection from having too many points of nondifferentiability because the function can't jump around too much). But, f does have infinitely many points lacking differentiability [19], and this set includes more than just the corners of the stairsteps.

Finally, a function known as the everywhere prickly curve, which is virtually impossible to graph, is defined by means of an infinite sum. First $f_1(x)$ is defined in agreement with the graph in Fig. 5-4 and then inductively defined as

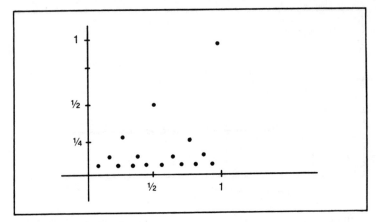

Fig. 5-2. The ruler function.

$$f_n(x) = 4^{1-n} f_1 (4^{n-1} x).$$

The everywhere prickly function f is the sum

$$f(x) = \sum_{n=1}^{\infty} f_n(x).$$

This function happens to be continuous everywhere (on the entire real line), but is nowhere monotonic (which means it is constantly

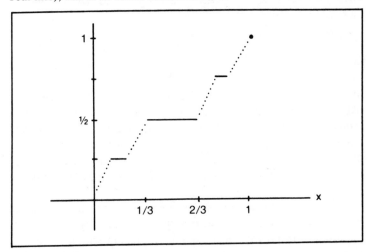

Fig. 5-3. The Cantor function.

210

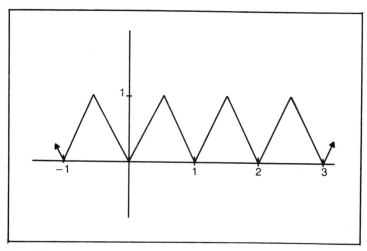

Fig. 5-4. Initial construction of the everywhere prickly curve.

jumping up and down). As a consequence, f is nowhere differentiable. Likewise, it is not of bounded variation nor absolutely continuous.

It is very difficult, if not impossible, to adequately work with these functions on a computer. Sometimes in analysis you just can't tolerate any roundoff error; this could easily change a number from irrational to rational or cause an infinite sum to be terminated after many steps. These changes might have a profound influence on the outcome.

All of the above functions were defined on a continuum of values. One last function, defined on the set of whole numbers, known as the Ackermann function, A(n), is a function that is sometimes studied in an advanced computer science course, usually using ALGOL, because A is by definition a recursive function, and ALGOL incorporates the recursion features. This function serves as an example of a nonprimitive recursive computable function. The function values grow so large (you'll see numbers bigger than any you've ever seen in your life—they will make a googolplex look like peanuts) that one is unable to give the explicit value for more than the first three or four function values. The languages BASIC and FORTRAN are literally of no help in dealing with the Ackermann function. A can be defined by first defining a set of functions f_0, f_1, f_2, ••• each having two independent variables, x and y, using the following double recursion formula:

$$f_0(x,y) = x + y$$

$$f_{n+1}(0,y) = \begin{cases} 0 & \text{if } n = 0 \\ 1 & \text{if } n = 1 \\ y & \text{if } n > 1 \end{cases}$$

$$f_{n+1}(x+1,y) = f_n(f_{n+1}(x,y),y).$$

This definition yields for the four functions, f_0, f_1, f_2, f_3, the values

$$f_0(x,y) = x + y$$
$$f_1(x,y) = xy$$
$$f_2(x,y) = y^x$$

$$f_3(x,y) = y^{y^{\cdot^{\cdot^{\cdot^{y}}}}} \qquad (1 + x \text{ occurrences of } y).$$

You can see a progression of the hierarchy of operations: first addition, then multiplication, then exponentiation, and then repeated exponentiation. The remaining functions, f_4, f_5, •••, are all of this last type, with repeated exponentiation of higher and higher degrees. You can then define A(n) to be equal to $f_n(n,n)$. Thus, A(0) = 0, A(1) = 1 + 1 = 2, A(2) = 2•2 = 4, and A(3) = $f_3(3,3)$ which must equal

$$3^{3^{3^{3}}}.$$

This value, A(3), is a tremendously large number containing roughly 3.63×10^{12} digits and having an approximate value of

$$10^{10^{10^{12.56}}}.$$

But, you haven't seen anything yet! The value of A(4) is of the form (also repeated exponentiation)

$$4^{4^{4^{\cdot^{\cdot^{\cdot^{4}}}}}}$$

where the digit 4 occurs a total of C times. The value of C is itself expressed as a repeated exponential,

$$C = 1 + 4^{4^{4^{\cdot^{\cdot^{\cdot^{4}}}}}}$$

where the digit 4 occurs B times; and B is of the form

$$B = 1 + 4^{4^{4^{\cdot^{\cdot^{\cdot^{4}}}}}}$$

where the 4 occurs AA times, with

$$AA = 1 + 4^{4^{4^{4^{4^{4}}}}}$$

It would prove a tough enough task determining the magnitude of AA without ever getting to A(4).

Now, obviously, the functions that are used in the remainder of this chapter are not of this complexity, especially since you want to be able to perform some computer calculations. Some of them may appear to be a little nasty, but those would probably fall under the somewhat nice category, if not even in the nice category.

SEQUENCES

The notion of a sequence is of fundamental importance in mathematics, particularly in those areas closely related to analysis. Recall that a sequence, say $\{a_n\}$, is merely a collection of numbers, but the collection is ordered, which means there is a definite first term, a_1, then a second term, a_2, and so on:

$$a_1, a_2, a_3, \cdots .$$

There are several important questions that mathematicians ask themselves concerning each given sequence. Do the terms follow a particular pattern, and is there a general formula for the n-th term? Do the terms tend to increase, decrease, or oscillate in a predictable pattern? And do the terms tend to level off in value and approach some definite number, and if so, what is this limiting value? To a mathematician, a sequence is merely a function defined on the set N of natural numbers, so $f(n) = a_n$ with $n \in N$. Graphing the sequence $\{a_n\}$ amounts to graphing f, which would appear as a succession of isolated points in the right-half of the Cartesian plane, as shown in Fig. 5-5. Consequently, knowledge of the sequence will imply certain results about the behavior of the function; and vice versa, knowledge of f will aid in describing $\{a_n\}$.

For example, suppose the sequence $\{a_n\}$ has for its first seven terms

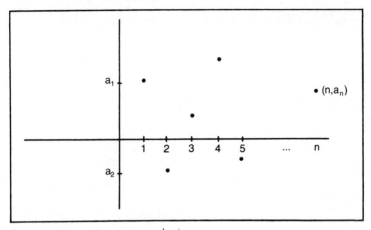

Fig. 5-5. Graph of the sequence $\{a_n\}$.

$$-3, \; -3, \; -1, \; 3, \; 9, \; 17, \; 27$$

and you would like to predict, with some authority, what the next few terms might be. It might be helpful to examine the graph of the sequence shown in Fig. 5-6.

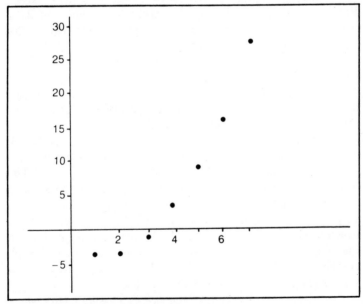

Fig. 5-6. The graph of a particular sequence.

Connecting the points with a smooth curve might suggest that the points lie on a polynomial path, possibly a parabolic path. If this were the case, there would be some polynomial of degree 2, say $ax^2 + bx + c$, for which

$$f(n) = an^2 + bn + c.$$

Substituting values from three of the points ($(1,-3)$, $(3,-1)$, $(5,9)$) into this equation gives the system,

$$-3 = a + b + c$$
$$-1 = 9a + 3b + c$$
$$9 = 25a + 5b + c$$

from which it follows that $a = 1$, $b = -3$, $c = -1$. It is possible then that the sequence follows the particular pattern described by

$$a_n = n^2 - 3n - 1.$$

Checking this out with some of the other points, $(2,-3)$, $(4,3)$, and $(6,17)$, would seem to establish this as fact. Thus, you could give rationale to declaring that the next term ($n = 8$) in the sequence if $a_8 = 39$.

Now suppose that there is a definite formula describing the terms in sequence, say for example,

$$a_n = [3 + n \cdot \sin(n)]$$

where the brackets denote the greatest integer function. The first few terms in this sequence are 3, 4, 3, -1, -2, 1, 7, 10, 6, -3. You might question whether or not the terms continue to oscillate between positive and negative values, and if they do, whether or not they are predictable. Are there positive values which get exceedingly large? Are there negative values which get exceedingly small? Is there an n for which $a_n = 0$; $a_n = 13472$?

Consider the sequence whose general term is given by

$$a_n = (1/n)^{1/n}.$$

The first few values are approximately $a_1 = 1$, $a_2 = (1/2)^{1/2} = .707$, $a_3 = .693$, $a_4 = .7\ 07$, $a_5 = .724$ and $a_6 = .742$. Proceeding further you find $a_{10} = .794$, $a_{20} = .861$, $a_{30} = .893$, $a_{40} = .912$ and $a_{50} = .925$. Is there any kind of noticeable pattern? It does appear that the terms continue to increase for all n greater than 3. Algebraically this would follow if you could establish $a_n < a_{n+1}$, or equivalently,

$$\left(\frac{1}{n}\right)^{1/n} < \left(\frac{1}{n+1}\right)^{1/(n+1)}$$

which simplifies to

$$(n+1)^n < n^{n+1}.$$

This is a very interesting inequality, and supposedly it holds true for all $n \geq 3$, but you haven't verified it yet! One approach is to consider the real valued function $f(x) = x^x$, defined for positive x, and to examine this curve on the interval [0, 1/3]. Since the derivative of f is

$$f'(x) = x^x [1 + \ln(x)]$$

then $f'(x) < 0$ when $\ln(x) < -1$, and $f'(x) > 0$ when $\ln(x) > -1$, since $\ln(x) = -1$ when $x = 1/e$. A sketch of f appears in Fig. 5-7.

Substituting 1/n in place of x, and letting $n \to \infty$ ($x \to 0^+$), you can see that $a_n = (1/n)^{1/n}$ must continue to increase. In fact it does tend to converge to a limiting value of one.

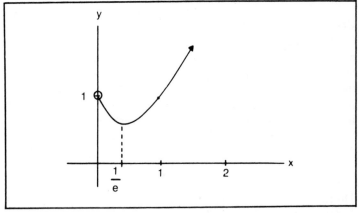

Fig. 5-7. Graph of $y = x^x$.

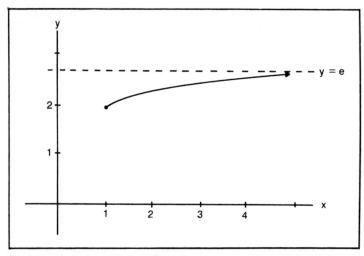

Fig. 5-8. Graph of $y = (1 + 1/x)^x$, $x \geqslant 1$.

Finally, a class of sequences that converges to some well known numbers will be examined. The first sequence is given by

$$a_n = (1 + 1/n)^n$$

which is an increasing sequence that converges to $e = 2.71828 \cdots$, the base of the natural logarithm. The graph of the corresponding function $y = (1 + 1/x)^x$, $x \geqslant 1$, bears out these two characteristics as shown in Fig. 5-8. Now, oddly enough, the similar looking sequence $\{b_n\}$ given by

$$b_n = (1 + 1/n)^{n+1}$$

also converges to e, but the sequence is entirely decreasing, as shown by the graph of the function $(1 + 1/x)^{x+1}$ in Fig. 5-9. Merely by changing the exponent from n to n+1 has the dramatic effect of completely reversing the ordering of the terms. Now consider a similar type of sequence that has an exponent that lies between n and n+1. To this end c_n is defined by

$$c_n + (1 + 1/n)^{n+\alpha}$$

where α is some given constant satisfying $0 < \alpha < 1$. For instance, suppose $\alpha = .49$; the first few terms in the sequence are listed in

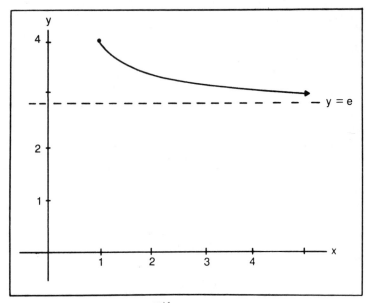

Fig. 5-9. Graph of $y = (1 + 1/x)^{x+1}$, $x \geq 1$.

Table 5-1. It appears that the sequence is decreasing, but because of the face that $a_n < c_n < b_n$, and both $\{a_n\}$ and $\{b_n\}$ converge to $e = 2.71828\cdots$, $\{c_n\}$ must also converge to e. Since c_{10} is less than e, sooner or later the sequence must start to increase. This turnabout takes place at $n = 17$ because $a_{16} > a_{17}$ and $a_{17} < a_{18} < a_{19} < \cdots$. From

Table 5-1. Some Values of c_n.

n	$c_n = (1 + 1/n)^{n + .49}$
1	2.8089
2	2.7445
3	2.7292
4	2.7235
5	2.7208
6	2.7195
7	2.7187
8	2.7182
9	2.7179
10	2.7178

a function viewpoint, the graph of $f(x) = (1 + 1/x)^{x+.49}$ will be decreasing for positive x up until approximately $x = 17$, then the function will slowly climb upwards and tend to an asymptotic value of e as shown in Fig. 5-10. This property possessed by $\{c_n\}$, when it initially decreases and then eventually increases, is characteristic of many sequences of the form $(1 + 1/n)^{n+\alpha}$, but certainly not of all. The property holds [20] only for α in the interval $(.409, .5)$. For α smaller than .409 the sequence is always increasing and convergent to e; while if α is greater than or equal to .5, the sequence decreases to e.

This data could have been easily compiled by running a program that examined the monotonicity (increasing or decreasing) of the sequence

$$(1 + 1/n)^{n+\alpha}$$

as α varied from 0 to 1 in some small increment. Now that you are comfortable with the above circumstances, turn your attention to a generalized problem that may provide some interesting results.

For this, consider the sequence $\{d_n\}$ given by

$$d_n = (1 + A/n)^{n+\alpha}$$

where both A and α are given constants. Ask the same questions about $\{d_n\}$ as you did about $\{c_n\}$, namely is this sequence increasing

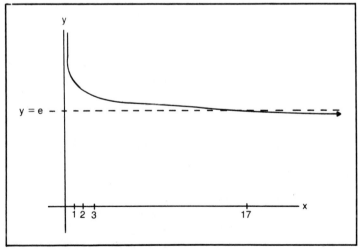

Fig. 5-10. Graph of $y = (1 + 1/x)^{x+.49}$.

(for what values of A, α), decreasing, or doing some of both? If the sequence converges, to what value does it tend?

You may wish to run A, α through a loop, assuming, for example, the values from -5 to 5, with increments of $h = 1/10$. For each such choice of A, α, you will compute d_1, d_2, d_3, $\cdots$, d_M with the computations terminated as soon as two consecutive terms differ from one another by less than a given tolerance, ($|d_M - d_{M-1}| <$ TOLER), say TOLER $= .00001$. The tolerance will need to be quite small, as exemplified by the previous sequence $\{c_n\}$ with $\alpha = .49$. Moreover, $\{d_n\}$ will either be entirely increasing, or entirely decreasing, or a single combination of the two (that is, increasing for $d_1 < d_2 < \cdots < d_k$ and then decreasing for the remainder $d_k > d_{k+1} > \cdots$). It will be important to maintain the value of this turn-about point k.

You can put the elements in the sequence in an array, D(n). In this case a dimension statement will be needed, for example, DIM D(1000). Alternatively you can view the sequence as a function. In this case a function definition statement will be needed;

$$\text{DEF FND}(x) = (1 + A/x)**(x + \alpha).$$

Consider the flowchart shown in Fig. 5-11, which describes a program that computes values of $\{d_n\}$, determines whether it is increasing or decreasing, stores the value of the turn-about point, and estimates the value the sequence converges to.

Exercises

1. Write a program that determines the monotonicity of the sequence $(1 + A/n)^{n+\alpha}$ and what its limiting value is. Compile this data in a table for various A, α. Can you locate a pattern? When will the sequence be entirely increasing? Test your program for the two specific sequences when $\alpha = -.6$, $A = -1$, and when $\alpha = 1.41$, $A = \pi$.
2. The sequence $(1 + \sqrt{2}/n)^{\pi n + e}$ converges to what value?

DIFFERENTIATION

The most fundamental operation that a student learns in elementary calculus is that of differentiation. This is an operation that is applied to a single real-valued function, and which then produces (essentially) a new function (only in rare cases are the two functions identical). You can perhaps think of the operation as a

220

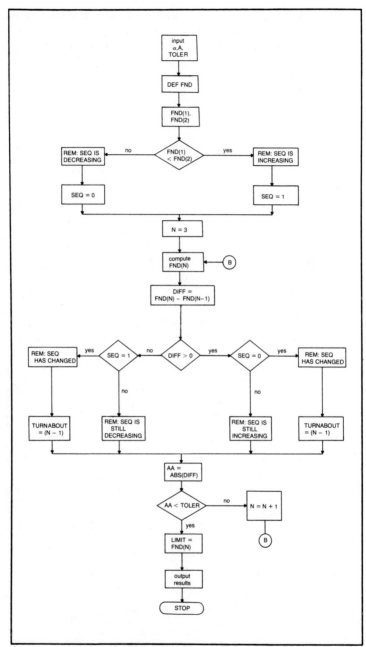

Fig. 5-11. Flowchart describing the behavior of $(1 + A/n)^{n+\alpha}$.

machine. When a function f is fed into the machine, the result output is a new function f′, called the derivative of f as shown in Fig. 5-12. If the function f is a function of the single real variable x, f(x), the differentiation operator is customarily denoted by

$$\frac{d}{dx}$$

and the effect of applying this to f can be symbolized by

$$\frac{d}{dx}(f) = f'.$$

Some examples to illustrate the process include:

1. $f(x) = x^2 + 3,$ $\qquad$ $f'(x) = 2x$
2. $f(x) = \sin(1/x),$ $\qquad$ $f'(x) = -\cos(1/x)/x^2$
3. $f(x) = e^{2x},$ $\qquad$ $f'(x) = 2e^{2x}.$

    This brings you to the stage of determining the explicit nature of f′. How does one ascertain the equation for f′? The answer to this

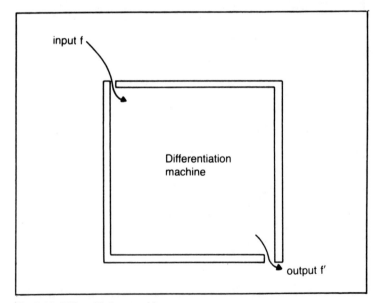

Fig. 5-12. Differentiation machine.

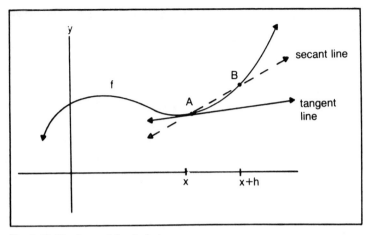

Fig. 5-13. The tangent line to f at x.

lies in a thorough understanding of the definition of differentiation, which is now stated.

Definition: Given a real-valued function f(x), the derivative f'(x) is given by the limit,

$$f'(x) = \lim_{h \to 0} \left[\frac{f(x+h) - f(x)}{h} \right]$$

whenever the limit exists.

For those that are unfamiliar with this expression and who question the meaning of the difference quotient $[f(x+h) - f(x)]/h$, its usage can be easily explained if you examine the geometry involved. Consider the function f, which is at least continuous, and two arbitrary points A(x,f(x)), B(x+h,f(x+h)) that belong to the graph of f. The secant line connecting A and B has a slope equal to the above difference quotient. Then, as the value of h tends to diminish to zero (h → 0), the point B will "slide down the curve" and approach point A; this is where the continuity of f is crucial. In so doing, the secant line gradually approaches a limiting position, a position reserved for what is called the "tangent line to the function f at x," as shown in Fig. 5-13. As a consequence, f'(x) represents the slope of the tangent line!

The derivative of a function is nothing more than an expression for the slope of the tangent line to the curve at the general point (x,f(x)). Slopes of specific tangent lines can be computed by merely substituting the particular value of x into f'(x). Thus, because f'(x) = 2x when f(x) = x², the slope of the tangent line to f at the point (3,9) is f'(3) = 2•3 = 6.

The tangent line is so named because the line is to intersect the graph of the function at just one point, (x,f(x)). It may very well be, and it usually is, that the line intersects the graph at some other point, but this other point is a considerable distance away from (x,f(x)). The tangent line shown in Fig. 5-14 would appear to intersect the graph at both (x,f(x)) and the distant point P. There is nothing wrong with this happening; a tangent line can actually intersect the graph in many points. The restriction, though, is that there must be some small region around (x,f(x)) which contains no other points of intersection. A typical region R is shown in Fig. 5-15.

Unless you are familiar with standard mathematical techniques for determining f', the process of evaluating some derivatives will prove quite troublesome. It is an easier matter to compute derivatives at specific points, and here is where the computer becomes useful. If, for example, you wish to approximate the derivative of f(x) = x² at the point (3,9), you must resort to the definition that considers the limiting value of the slopes of secant lines,

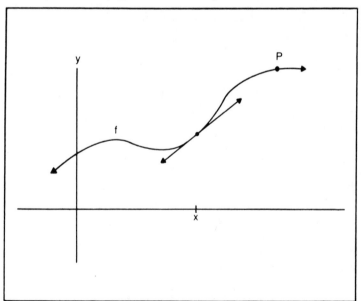

Fig. 5-14. Tangent line intersecting graph at two points.

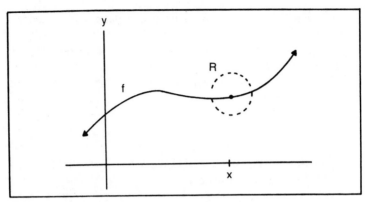

Fig. 5-15. Region containing single point of intersection.

$$\lim_{h \to 0} \frac{f(x+h) - f(x)}{h}$$

At $x = 3$ you have,

$$\lim_{h \to 0} \frac{f(3+h) - f(3)}{h}$$

You can choose successively smaller and smaller values for h, say h = 1, .5, .25, .125, .0625, ••• and have the computer evaluate the difference quotient. For example, if h = 1, the difference quotient equals $[f(4) - f(3)]/1 = (16 - 9)/1 = 7$. Further values are included in Table 5-2.

Table 5-2. Evaluating Various Different Quotients for $y = x^2$.

h	$\dfrac{f(3+h) - f(3)}{h}$
1	7
.5	6.5
.25	6.25
.125	6.125
.0625	6.0625
.03125	6.03125
.015625	6.015625
.0078125	6.0078125

You would want to terminate these calculations when successive values of the difference quotient are reasonably close together. This tolerance, call it ϵ, can be prescribed at the outset; possibly $\epsilon =$ .01. If the values of the difference quotients are stored in an array, say DIFFQT, then you could terminate the calculations and pass out of the loop as soon as

$$| \, \text{DIFFQT(i+1)} - \text{DIFFQT(i)} \, | < \epsilon.$$

The approximate value of the derivative of f(x) at x would be given by the value of DIFFQT(i+1); thus

$$f'(x) = \text{DIFFQT(i+1)}.$$

In the above example, with $f(x) = x^2$, the derivative at $x = 3$ would be approximately $f'(3) = 6.0078125$.

A sample program that could be used to compute the approximate derivative of $f(x) = x^2$ at an arbitrary point (a,f(a)) is listed below.

Program

```
100   REM APPROXIMATE DERIVATIVE OF Y = X² AT X
      = A
110   DIM DIFFQT(50)
120   INPUT ε
130   REM ε IS THE TOLERANCE FOR SUCCESSIVE
      DIFFERENCE QUOTIENTS
140   DEF FNA(X) = X*X
150   INPUT A
160   H = 1
170   DIFFQT(1) = (FNA(A+H) − FNA(A))/H
180   FOR I = 2 TO 50
190       POWER = H/2**(I−1)
200           DIFFQT(I) = (FNA(A + POWER) − FNA(A))
              /POWER
210       IF ABS(DIFFQT(I) − DIFFQT(I−1)) < ε GO TO
          240
220   NEXT I
230   PRINT "NO SOLUTION" GO TO 250
240   PRINT "THE DERIVATIVE AT X = "; A; "IS AP-
      PROXIMATELY"; DIFFQT(I)
```

250 STOP
260 END

It is important to mention that the existence of a derivative at a point implies a certain degree of smoothness of the function, smoothness at the point. This concept is quite technical and demands a greater understanding of differentiation. Suffice it to say that a function that is discontinuous at a point x = a does not have the smoothness there that a continuous function has. Likewise, a continuous function that fails to have a derivative at x = a is not as smooth as one that does have a derivative [70]. Such is the case with f(x) = |x|, which is continuous but not differentiable at x = 0. A function that is differentiable at a point may not be as smooth as another differentiable function. Why is this? Because of the concept of higher order derivatives.

A second order derivative of a function f, denoted by f ′ ′, is simply the derivative of the function f′. Likewise the third derivative of f, denoted by f′ ′ ′, is the derivative of f′ ′. In general the n-th order derivative $f^{(n)}$ is given by

$$f^{(n)} = \frac{d}{dx} (f^{(n-1)}).$$

Mathematicians tend to equate the degree of smoothness of a function with the highest order derivative that the function possesses. Functions exist that have as many derivatives (including infinitely many) as anyone would like. The function $f(x) = x^{n + 1/3}$ has n derivatives that exist and are defined for all real numbers, but the (n+1)st derivative fails to exist since $f^{(n)}$ is not differentiable at x = 0.

The discussion will be limited to the second derivative for the remainder of this section. The value of f′ ′ at the point x is given by the same limit definition discussed earlier, namely

$$f' ' (x) = \lim_{h \to 0} \frac{f' (x+h) - f'(x)}{h}$$

For example, with $f(x) = x^2$, and f′(x) = 2x, the second derivative evaluated at x = 3 would be given by,

$$f' ' (3) = \lim_{h \to 0} \frac{f' (3+h) - f' (3)}{h}$$

227

$$= \lim_{h \to 0} \frac{2(3+h) - 6}{h}$$

$$= \lim_{h \to 0} 2$$

$$= 2.$$

A more complicated example is $f(x) = \sin(1/x)$. Then, since

$$f'(x) = \frac{-\cos(1/x)}{x^2}$$

the value of f'' at $x = 1$ would be defined by,

$$f''(1) = \lim_{h \to 0} \frac{f'(1+h) - f'(1)}{h}$$

$$= \lim_{h \to 0} \frac{\dfrac{-\cos(1/(1+h))}{(1+h)^2} - \dfrac{-\cos(1)}{1}}{h}$$

You could, if unfamiliar with more advanced techniques, program the computer to estimate this limit using the method discussed earlier involving difference quotients. For the time being, this method of approximating f'' at a point will be called Method A. This is to distinguish it from Method B, which is similar to Method A in one strong aspect, namely that it too depends on evaluating the limit of a particular set of values. But, the big difference is that knowledge of f' is not needed for Method B; it relies only on the values of the function f. The defining equation [67] for Method B is given by,

$$f''(x) = \lim_{h \to 0} \frac{f(x+h) + f(x-h) - 2f(x)}{h^2}$$

Exercises

1. Suppose $f(x) = \sin(1/x)$ and $g(x) = x^x$. It follows that $f'(x) = (-1/x^2)\cos(1/x)$ and $g'(x) = x^x(1 + \ln(x))$. Write a program that

228

approximates the values of f′′(2) and g′′(2) by employing both methods A and B for each derivative. Terminate your limit approximations when the quotients are within $\epsilon = .01$ of each other. Record the number of iterations that were necessary to achieve this, and print them out. In each case, use the values for h of h = 1, 1/2, 1/4, 1/8, •••.

2. Give an example of a function f which is three-times differentiable at x = 1, but $f^{(4)}(1)$ does not exist.

DIRECTIONAL DERIVATIVES

Although a great variety of situations can be described by using functions of one real variable, f(x), many others call for the use of several variables. For example, the monthly profit of an ice cream store, P, may be a function of the number of employees x and the number of hours h per day of direct sunlight; the demand D for beef may be a function of the price P of beef and also the prices Q, R, and S of pork, mutton and fish; the volume V of a rectangular box depends upon the height h, the length L, and the width w. These three functions could be denoted using mathematical notation by,

1. P = f(x,h)
2. D = f(P,Q,R,S)
3. V = f(h,L,w).

The remainder of this section will be devoted to functions of two variables, f(x,y), although some of the concepts will generalize quite easily to functions of many variables. The significant new element in dealing with two variables is that the graphs of functions are surfaces in three-dimensional space, and this presents some visual problems. The notion of the derivative of a function, f′(x), as was discussed in the previous section will now be generalized and extended. Recall that this derivative is defined by the limit

$$f'(x) = \lim_{h \to 0} \frac{f(x+h) - f(x)}{h} .$$

What then would be the logical means for defining the derivative of f(x,y)? Since the derivative takes into account the independent variable x, it is only natural to think that f(x,y) has two derivatives, one that considers x as the independent variable and the other that considers y as the independent variable.

If you consider x as the independent variable; then y is treated as a constant. The derivative of f(x,y) is then called the partial derivative of f with respect to x. This is denoted by,

$$\frac{\partial f}{\partial x}$$

Correspondingly, the partial derivative with respect to y (x is treated as a constant) is denoted by

$$\frac{\partial f}{\partial y}$$

To illustrate, suppose $f(x,y) = 2x^2y^3$. The partial derivative with respect to x is given by

$$\frac{\partial f}{\partial x}(x,y) = 4xy^3.$$

Keep in mind that y is constant. The partial derivative with respect to y (x is constant) is

$$\frac{f}{y}(x,y) = 6x^2y^2.$$

Several other examples would include,

1. $f(x,y) = x/y,$ $\quad \frac{\partial f}{\partial x}(x,y) = 1/y,$ $\quad \frac{\partial f}{\partial y}(x,y) = -x/y^2$

2. $f(x,y) = x\sin(y),$ $\frac{\partial f}{\partial x}(x,y) = \sin(y),$ $\frac{\partial f}{\partial y}(x,y) = x\cos(y).$

It is important to realize the alternate role played here with one of the variables acting as the independent variable and the other acting as constant. You can thus formulate a definition for these partial derivatives as limits of difference quotients:

1. $\quad \frac{\partial f}{\partial x}(x,y) \quad = \quad \lim_{h \to 0} \frac{f(x+h,y) - f(x,y)}{h}$

$$2. \ \frac{\partial f}{\partial y} \ (x,y) \ = \ \lim_{h \to 0} \ \frac{f(x,y+h) - f(x,y)}{h}$$

The geometric interpretation of partial derivatives is quite similar to the derivative of a function of a single variable. In the latter case, the derivative was the slope of the line tangent to the curve at the specific point in question. With partials, say $\frac{\partial f}{\partial x}(a,b)$, this derivative is also the slope of a line, the line that is tangent to the curve determined by the intersection of the surface $z = f(x,y)$ and the plane $y = b$; the point of tangency is the point determined by $x = a$, $y = b$. Similarly, $\frac{\partial f}{\partial y} \ (a,b)$ is the slope of the line tangent to the curve determined by the intersection of the three-dimensional surface $z = f(x,y)$ and the plane $x = a$, at the same specific point.

The slopes of tangent lines offer one way to interpret derivatives. Another way is to recognize that a derivative indicates a particular rate of change, actually the rate at which the dependent variable (the function value) changes per unit increase in the independent variable, at the specific moment when the variables assume the values indicated at the point. If you consider $f(x,y) = 2x^2y^3$, and use the point P where $x = 2$ and $y = 1/2$ (P $= (2,1/2,1)$) then

$$\frac{\partial f}{\partial x} \ (2,1/2) = 4(2)(1/2)^3 = 1.$$

This derivative of 1 represents a 1 unit increase in the function value per unit increase in x, with y remaining constant at the instant when $x = 2$ and $y = 1/2$. You could say that the function change will be approximately equal to the change in x as x changes slightly from its value of 2, but y must stay equal to 1/2. Similarly, since

$$\frac{\partial f}{\partial y} \ (2,1/2) = 6(4)(1/2)^2 = 6$$

the function value changes by approximately 6 times as much as y changes, when y changes from 1/2 to something close to 1/2, and x must remain equal to 2.

Knowing these rates of change is important, but you should

231

keep in mind that they are only the rates of change of f as the point (a,b,f(a,b)) moves parallel to the x or y-axis. In an attempt to better understand the nature and character of the function, you need to know its rate of change as the specific point moves in any direction along the surface defined by f. To this end the directional derivative of f at a point (a,b,f(a,b)) is defined as in the direction of the unit vector $u_1 i + u_2 j = \; <u_1,u_2>$. See Fig. 5-16.

Definition: If $\vec{u} = \; <u_1,u_2>$ is a unit vector, and f(x,y) is a function of two real variables, the derivative $D_u f(a,b)$ of f at the point (a,b,f(a,b)) in the direction of the unit vector u is given by,

$$D_u f(a,b) = \frac{\partial f}{\partial x}(a,b) \cdot u_1 + \frac{\partial f}{\partial y}(a,b) \cdot u_2.$$

To better understand this definition several examples shall be worked through. If you let $f(x,y) = 2x^2 y^3$, (a,b) = (2,1/2) and $u = \; <u_1,u_2> = \; <3/5,4/5>$, then the derivative of f at (2,1/2,1) in the direction of 3i/5 + 4j/5 is

$$D_u f(2,1/2) = \frac{\partial f}{\partial x}(2,1/2) \cdot (3/5) + \frac{\partial f}{\partial y}(2,1/2) \cdot (4/5)$$
$$= 1 \cdot (3/5) + 6 \cdot (4/5)$$
$$= 27/5.$$

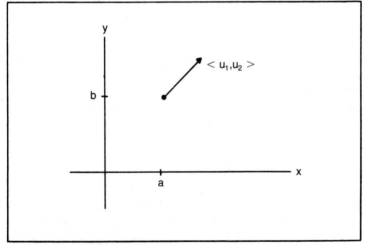

Fig. 5-16. The unit vector $<u_1,u_2>$.

On the other hand, if you wanted the derivative of $f(x,y) = 2x^2y^3$ at $(2,1/2,1)$ in the direction toward the point $(3,1,18)$, you would first have to compute the unit vector for that direction, which would be

$$\vec{u} = \frac{(3-2)i + (1 - 1/2)j}{\sqrt{(3-2)^2 + (1 - 1/2)^2}}$$

$$= \frac{2i + j}{\sqrt{5}}$$

Then the derivative would be

$$D_u f(2,1/2) = \frac{\partial f}{\partial x}(2,1/2){\cdot}(2/\sqrt{5}) + \frac{\partial f}{\partial y}(2,1/2){\cdot}(1/\sqrt{5})$$

$$= 1{\cdot}(2/\sqrt{5}) + 6{\cdot}(1/\sqrt{5})$$

$$= 8/\sqrt{5}.$$

The interesting thing here is that as you proceed from the point $(a,b,f(a,b))$ and head in any direction possible, the set of all possible values for the directional derivative is an infinite set because there are infinitely many directions. Yet this is an infinite set that has a largest element! In other words, given a point $(a,b,f(a,b))$ on the surface formed by f, the set

$$\{D_u f(a,b): \vec{u} = \ <u_1,u_2>\ , -1 \leqslant u_1 \leqslant 1, -1 \leqslant u_2 \leqslant 1, u_1^2 + u_2^2 = 1\}$$

contains a largest element.

At this point the computer program that searches through this set for the maximal element should be discussed. To start with, assume that you are given two pieces of information to work with: the function f and the values a and b. The program can be broken down into three steps:

1. approximate the partial derivative $\dfrac{\partial f}{\partial x}(a,b)$

2. approximate the partial derivative $\dfrac{\partial f}{\partial y}(a,b)$

3. run through the possible values for u_1 and u_2 and determine which one maximizes $\frac{\partial f}{\partial x}(a,b) \cdot u_1 + \frac{\partial f}{\partial y}(a,b) \cdot u_2$.

Both 1 and 2 can be programmed in the same manner. Since

$$\frac{\partial f}{\partial x}(a,b) = \lim_{h \to 0} \frac{f(a+h,b) - f(a,b)}{h}$$

this limit can be approximated by choosing small enough values for h until the difference between successive DIFFERENCE QUOTIENT values is less than some prescribed tolerance. This method is the same as the one discussed in the previous section on derivatives of functions of a single variable.

Because of the conditions on u_1 and u_2, these values correspond to the coordinates of the points on the unit circle as shown in Fig. 5-17. By choosing small enough increments, you could consider a large enough sample of points to adequately evaluate $D_u f(a,b)$. You can eliminate 75 percent of the possible values for u_1 and u_2 by selecting the proper quadrant that these values lie in. This follows because u_1 would have the same algebraic sign (plus or minus) as

$\frac{\partial f}{\partial x}(a,b)$, and u_2 would have the same sign as $\frac{\partial f}{\partial y}(a,b)$. For instance,

if $\frac{\partial f}{\partial x}(a,b) < 0$ and $\frac{\partial f}{\partial y}(a,b) > 0$ you would choose $u_1 < 0$ and

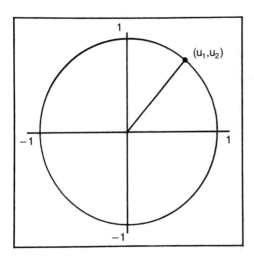

Fig. 5-17. Unit vectors correspond to points on the unit circle.

$u_2 > 0$, so these values would lie on the arc of the unit circle in quadrant 2. Consequently, you could consider 1000 of these different unit vectors $\vec{u} = <u_1, u_2>$ using the following program.

```
100   FOR I = 1 TO 1000
110       U1 = −1 + I/1000
120       U2 = SQRT(1 − U1*U1 )
130   NEXT I
```

Between statements 120 and 130 you would evaluate the directional derivative $D_u f(a,b)$ and compare this value with the largest value obtained so far. If this new value is larger, you should store the pertinent information, namely the derivative value and the vector u. If the value is not larger, proceed to the next point (u_1, u_2).

Exercises

1. Write a program that determines the approximate maximum directional derivative for the function $f(x,y) = 2x^2 \sin(y)$ at the specific point where $x = 2$, $y = \pi/4$.

2. Suppose you have a function f of three variables, $f(x,y,z)$. How could you define the directional derivative of f at $x = a$, $y = b$, $z = c$ in the direction of $x = d$, $y = e$, $z = g$?

FIXED POINTS

It is a fundamental question in mathematics to ask for the zeros of a function, that is to solve the equation $f(x) = 0$. Students are exposed to this same question as early as high school Algebra I, where they first learn how to solve linear equations, $ax + b = 0$, and then quadratic equations, $ax^2 + bx + c = 0$. In Trigonometry, students are able to solve $\sin(2x + 1) − \pi/4 = 0$; while in calculus they solve $e^{2x} − 1 = .4(e^{2x} + 1)$.

When it becomes difficult if not impossible to solve an equation, the best alternative is to resort to different methods. Approximation methods have to be tried, but, you should ascertain ahead of time whether the equation actually has a solution! There is no sense in fumbling around in the dark searching for a non-existent solution.

Many methods are available for approximating the zeros to certain types of functions under particular, though not unreasonable, circumstances. The remainder of this section is devoted to one such method, a method that involves the concept of fixed points.

Simply stated, a point p is said to be a fixed point of a function f if f(p) = p. This means that the function maps the value p into itself; the element p remains fixed under the influence of the function. The geometric interpretation would be that the graph of y = f(x) intersects the line y = x at the point (p,p). Of course it may intersect the line in other points as well, but this would imply that f has other fixed points in addition to p. See Fig. 5-18.

A couple of examples may be appropriate here. The function f(x) = x^3 has three fixed points at p = 1, 0, and −1; while the function f(x) = x − sin(πx), defined on [0,1], has two fixed points at p = 0, and 1. What are the fixed points for f(x) = (x^2 − 1)/3 defined on [2,4]?

Now, of course, a function does not have to have any fixed points, and this is the case many times. The elementary functions y = e^x and y = ln(x) do not have any fixed points. The same is true for the quadratic y = x^2 + bx + 1 whenever b ≤ 1 or b ≥ 3. It's convenient, though, to be able to determine whether or not a function has a fixed point without explicitly determining the point. This is because it may either be a lot of work to find the point, or the work could all be in vain if no such point exists. Several conditions that indicate that a function definitely has a fixed point will now be presented.

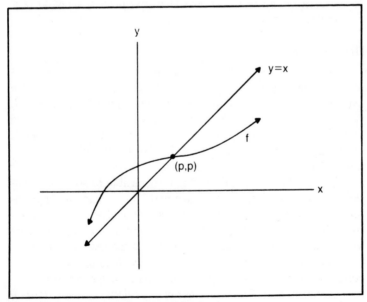

Fig. 5-18. Geometric interpretation of fixed points.

The first condition assumes that f is continuous on the closed interval [a,b]. If the range of f (i.e., the collection of all values f(x), as x ranges over [a,b]) is a subset of the same interval [a,b], it must follow that f has a fixed point somewhere in [a,b]. As an illustration, consider the linear function f given by f(x) = x/4 − 1. On the interval [−4,8], the function is continuous, and its range is precisely the interval [−2,1]. Since this interval is contained in [−4,8] you know that f must have a fixed point p where −4 < p < 8. This situation is depicted in Fig. 5-19. The exact value of p can be found by solving the equation p/4 − 1 = p, which yields p = −4/3. The point is that you knew for sure that a fixed point existed because f was continuous over [a,b], and the range was contained in the domain. But why does this method work?

First, one of the following is true: f(a) = a, or f(b) = b, or f(a) > a and f(b) < b. If either of the two equalities hold, a fixed point is already located. Assume then that f(a) > a and f(b) < b. If you define a new function F using the rule F(x) = f(x) − x, it follows that F is also continuous on [a,b] and that F(a) = f(a) − a > 0 and F(b) < 0. Consequently, the Intermediate Value Theorem asserts the existence of a number c located between a and b for which F(c) = 0. This c is the fixed point of f.

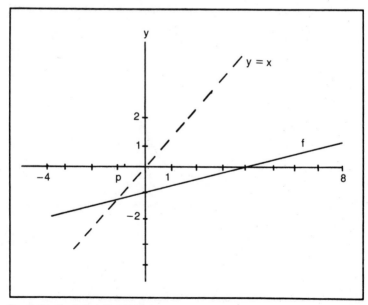

Fig. 5-19. Illustrating fixed points.

In order to recognize that there is only one fixed point as opposed to several, you need an additional restriction on f. Thus, if f is differentiable (a condition on the smoothness of f) on the open interval (a,b), and if the derivative values f'(x) are not too large, then f will have precisely one fixed point. More precisely, if there is some number k < 1 for which the absolute value of the derivative satisfies

$$|f'(x)| \leq k$$

for all x in (a,b), f will have a unique fixed point. Having the derivatives all less than one forces two arbitrary function values $f(x_1)$ and $f(x_2)$ to be closer together than x_1 and x_2. This is crucial since in prohibits the graph of f from intersecting the graph of y = x in more than one place. Note that in Fig. 5-20, where there are two such intersection points, there must be values for f'(x) that exceed one, since one is the slope of y = x.

The proof of this uniqueness property follows from a result from introductory calculus. If you suppose, on the contrary, that there are two such fixed points, p_1 and p_2, then the Mean Value Theorem asserts the existence of a point γ in [a,b] for which

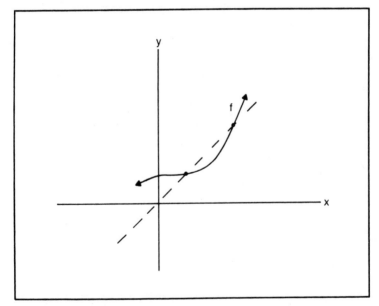

Fig. 5-20. A function with two fixed points.

$$\frac{f(p_2) - f(p_1)}{p_2 - p_1} = f'(\gamma)$$

but $f'(\gamma) \leqslant k < 1$, so this implies

$$1 = \frac{p_2 - p_1}{p_2 - p_1} = \frac{f(p_2) - f(p_1)}{p_2 - p_1} \leqslant k < 1$$

which is impossible. There must not be two fixed points with the above conditions on f.

These results are all fine and dandy, and most useful in discussing fixed points. The discussion, however, is incomplete without mentioning some way to obtain the value, exact or approximate, of the fixed point. One such method is that of functional iteration.

Table 5-3. An Iterative Search for a Fixed Point.

n	x_n
1	1
2	1.5
3	1.416667
4	1.414215
5	1.414163
6	1.414213
7	1.414213

In order to approximate the fixed point p, you should choose an initial value x_1 and then generate the sequence $\{x_n\}$ by $x_n = f(x_{n-1})$. This sequence will converge to p, $x_n \to p$, whenever f satisfies the previously stated conditions. To see this, consider $f(x) = x/2 + 1/x$ on the interval [1,3]. The function is continuous, its range is contained in [1,3]; and the absolute value of its derivative, $|f'(x)| = |1/2 - 1/x^2|$, is less than 1/2 for $1 < x < 3$. You therefore know that f has a single fixed point p somewhere between 1 and 3. Suppose you set your initial value x_1 equal to 1. The first few terms in $\{x_n\}$ are computed and listed in Table 5-3. It becomes apparent that the sequence values are converging to a value close to 1.4142, which would be an approximation for p. The actual value can be obtained by solving the equation

$$\frac{x}{2} + \frac{1}{x} = x$$

which yields $x = \sqrt{2}$.

How important is the choice of x_1 as a first approximate? Would the sequence $\{x_n\}$ still converge if x_1 were something other than one? For example, setting $x_1 = 2.5$, the sequence shown in Table 5-4 gives the following results, and you can see that again it apparently converges to $\sqrt{2}$. In fact, the convergence is quite rapid. In each case, the sequence converges to p, the fixed point. This leads us to mention the final two results.

Result 1. If f is continuous and if the sequence $\{x_n\}$ converges (where $x_{n+1} = f(x_n)$, and x_1 is chosen arbitrarily), its limit is a fixed point of f.

Result 2. If f is continuous, and its derivative satisfies $|f'(x)| \leq k < 1$, $\{x_n\}$ converges to a fixed point.

The proof of this latter result follows from an application of the Mean Value Theorem. To see this, arbitrarily select x_1, and then set $x_{n+1} = f(x_n)$ for $n = 2,3,\cdots$. Then, for any $n > 1$,

$$
\begin{aligned}
|x_{n+1} - x_n| &= |f(x_n) - f(x_{n-1})| \\
&= |f'(\gamma_n)|\, |x_n - x_{n-1}|) \quad \text{Mean Value Theorem.} \\
&\leq k \cdot |x_n - x_{n-1}| \\
&\leq k^2 |x_{n-1} - x_{n-2}| \\
&\quad\bullet \\
&\quad\bullet \\
&\quad\bullet \\
&\leq k^{n-1}|x_2 - x_1|.
\end{aligned}
$$

Since $k^{n-1} \to 0$, the terms in $\{x_n\}$ gradually get close to one another, and hence converge. If we set $p = \lim x_n$, then

$$f(p) = f(\lim x_n) = \lim f(x_n)$$

because f is continuous. Finally, because of the functional iteration scheme defining $\{x_n\}$ you have

n	x_n
1	2.5
2	1.65
3	1.4310606
4	1.4143127
5	1.4142136
6	1.4142136

Table 5-4. Searching for a Fixed Point with $x_1 = 2.5$.

$$p = \lim x_{n+1}$$
$$= \lim f(x_n)$$
$$= f(\lim x_n)$$
$$= f(p)$$

which shows that p is a fixed point. The proof of the first result follows a similar vein. What is important here is that the choice of x_1 has no bearing on determining a false value for the fixed point. Any value of x_1 will produce the correct p. Some values may approximate p sooner than others. In fact, one bound [13] for the speed on convergence states,

$$|x_n - p| \leq \frac{k^{n-1}}{1 - k} \; |x_2 - x_1|.$$

Thus, for the function given earlier by $f(x) = x/2 + 1/x$ on [1,3], you have $k = 1/2$, $p = \sqrt{2}$, $x_1 = 1$ and $x_2 = 1.5$. Then, for any $n > 1$, the convergence inequality would read

$$|x_n - \sqrt{2}| \leq \frac{(1/2)^{n-1}}{1 - .5} \; |.5| = (1/2)^{n-1}.$$

You could therefore determine ahead of time the maximum number of terms that would need to be computed before x_n would differ from p by a given tolerance.

How does all of this discussion on fixed points tie in with the introductory remarks on finding zeros to functions? Suppose you seek the zeros to $f(x) = x^3 + 3x - 7$. This problem is equivalent to seeking the fixed points of the function g, where $g(x) = (7 - x^3)/3$.

The computer program that you shall concern yourself with simply allows you to use the functional iterative scheme to estimate fixed points for particular functions. You are probably more aware now of the importance of fixed points and the intricate details surrounding them.

Exercises

1. The equation to be solved is $7 + 3x - \sin(x) = 0$. Equivalently, you need to solve $(\sin(x) - 7)/3 = x$. This means you need the fixed points for $g(x) = (\sin(x) - 7)/3$. Since $g(0) = -2.33$, and $g(-5) = -2.01$, it follows that the graph of g intersects $y = x$ somewhere between -5 and 0; thus $-5 < p$ and $p < 0$. Because

241

$|g'(x)| \leq 1/3$ (so $k = 1/3$), you know the functional iterative scheme of $x_{n+1} = g(x_n)$ will converge to the point p for any arbitrary x_1 between -5 and 0. Write a program that approximates p to the nearest thousandth. Use a variety of values for x_1, and compare the convergence rates. How large an interval [a,b] can you choose x_1 from so that $\{x_n\}$ will still converge to p?

2. The sequence $\{x_n\}$ defined above by $x_{n+1} = f(x_n)$ with $f(x) = x/2 + 1/x$ was shown to converge to $\sqrt{2}$ when $x_1 = 1$. What does $\{x_n\}$ converge to when $f(x) = x/3 + 1/x$? When $f(x) = x/k + 1/x$ with $k > 1$?

LOCATING ROOTS FOR NONLINEAR EQUATIONS

The study of equations is of basic importance in all of mathematics. This topic was alluded to in the previous section on fixed points, where you learned that if the conditions were right, a zero for the continuous function f can be found by determining the limit of the sequence $\{x_n\}$ where

$$x_{n+1} = f(x_n).$$

Unfortunately, the conditions are not often right! Other criteria need to be formulated to help locate the zeros in this case.

To this end consider the most elementary method, the bisection method. This method, like all the others that will be discussed, relies on the Intermediate-Value Property of continuous functions. Briefly, this property says that if f is continuous on the interval [a,b], and if $f(a) > 0$ with $f(b) < 0$ (or vice versa, $f(a) < 0$ and $f(b) > 0$), then the function must cross the x-axis somewhere between a and b; which means a zero p is between a and b as shown in Fig. 5-21.

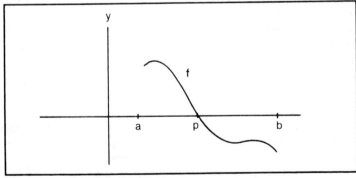

Fig. 5-21. Intermediate-value property.

Having determined (a computer can easily do this by checking, say, whether $f(n)f(n+1) < 0$ for $n = -100, -99, \cdots, 99, 100$) that the function changes signs from a to b, you make your first estimate x_1 for the value of p by setting x_1 equal to the midpoint of [a,b]; $x_1 = (a+b)/2$. Then compute $f(x_1)$, and check whether $f(x_1) = 0$ or is very close to zero (in which case you are done), $f(x_1) > 0$, or $f(x_1) < 0$. If $f(x) \neq 0$, you know that a zero (there actually may be several) lies in either the interval (a,x_1) or (x_1,b). By the Intermediate-Value Property the zero will lie in (a,x_1) if $f(a)f(x_1) < 0$ and will lie in (x_1,b) if $f(x_1)f(b) < 0$. Using the same sketch of f from Fig. 4-21, assume that $f(a) > 0$, $f(b) < 0$, and $f(x_1) < 0$. Thus, the interval (a,x_1) contains the zero as shown in Fig. 5-22.

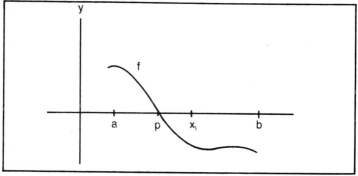

Fig. 5-22. The bisection method.

Your second estimate x_2 for p would then be the point you get by bisecting the interval (a,x_1); hence $x_2 = (a + x_1)/2$. Again, compute $f(x_2)$ and check its value. If it equals zero or is nearly zero, then you are done; otherwise, either $f(a)f(x_2) < 0$ or $f(x_2)f(x_1) < 0$, but not both. If, say, $f(x_2)f(x_1) < 0$ then p is located in the interval (x_2,x_1), and you can proceed by setting $x_3 = (x_2 + x_1)/2$. If you have not located a zero after n steps (so $f(x_i) \neq 0$ for $i = 1,2,\cdots,n$), you could terminate the program at one of several different stages: either when $|x_{n+1} - x_n|$ is sufficiently small or, which is usually preferred, when $|f(x_{n+1}) - f(x_n)|$ is small—smaller than some predescribed tolerance. The value x_{n+1} is then used as the approximate to p. This bisection method has its advantages, in that it is relatively simple to program, does not require much memory, and can be applied to any continuous function, given its equation. The drawback is that convergence can be quite slow, possibly with many computations required to approximate p with precision.

The bisection method is solely dependent upon the function values at x_1, x_2, x_3, •••. It fails to take into account the shape of the curve, and the rate of change of the function values. For instance, the three curves f_1, f_2, f_3 depicted in Fig. 5-23 are all defined on [0,1], with $f_i(0) = 1$ and $f_i(1) = -1$, but they all have quite a variance in how they curve downward and cross the x-axis. You need some iterative methods that account for this slope variance.

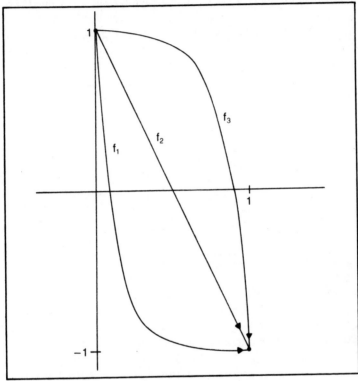

Fig. 5-23. Three functions depicting variance in slope.

The remaining five methods to be discussed all consider this feature. Furthermore, they are all of the form $x_{n+1} = F(x_n)$ where F is some function that accounts for the slope variance.

The Newton-Raphson iterative method [15] is one of the most commonly studied, for it makes direct use of the tangent line to f at the point $(x_n, f(x_n))$. Assuming f is differentiable, which it usually is, each successive approximation x_{n+1} to the zero p can be found by using the previous approximation x_n by means of the scheme,

$$x_{n+1} = x_n - \frac{f(x_n)}{f'(x_n)} \ .$$

The diagram that helps to interpret and explain this method is shown in Fig. 5-24. From the approximate x_n, you pass to the point $(x_n, f(x_n))$ on the graph of f. The tangent line to the curve at this point is extended until it intersects the x-axis; this intersection point is denoted by x_{n+1} and is found by subtracting the ratio $f(x_n)/f'(x_n)$ from x_n.

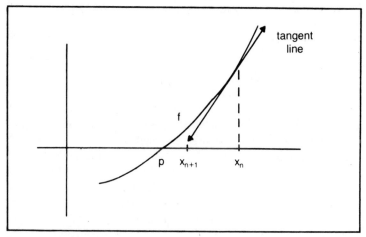

Fig. 5-24. Newton-Raphson method.

To illustrate Newton's method, consider the function $f(x) = x^3 + x - 1$, which has a zero in the interval (0,1). Since $f'(x) = 3x^2 + 1$ you have

$$x_{n+1} = x_n - \frac{x_n^3 + x_n - 1}{3x_n^2 + 1}$$

If you first set $x_1 = .5$, the next four values in the sequence are given in Table 5-5. For comparison sake the corresponding values from the bisection method are also listed. The value $x_5 = .6823278$ is close to the zero because $f(x_5) = -9.2 \times 10^{-9}$. You notice from this table how the Newton method is more efficient and rapid than the bisection method. A study of the error analysis of the two reveals that the Newton method converges quadratically (i.e., $|x_{n+1} -$

245

Table 5-5. Approximating the Zero to f(x) = x^3 + x − 1.

n	Newton-Raphson x_n	Bisection x_n
1	.5	.5
2	.7142857	.75
3	.6831797	.625
4	.6823284	.6875
5	.6823278	.65625

$p| \leq c|x_n - p|^2)$ while the bisection convergence is only liner.

Despite this advantageous property, the Newton method does have several drawbacks. First, it may be difficult to get an expression for f′(x), although this would probably only happen if f were of complicated form. Second, if both $f(x_n)$ and $f'(x_n)$ equal zero, or if just $f'(x_n)$ equals zero, the computer may spit out some nasty values. Third, the sequence $\{x_n\}$ may not converge! This is especially likely to happen if the value chosen for x_1 is not close enough to p. You should be leery of this and build a safeguard into the program to check the first few values x_1, x_2, x_3 and see that they are not heading in some divergent direction (what might be a good way to achieve this?).

A modified version of Newton's method calls for

$$x_{n+1} = x_n - \frac{mf(x_n)}{f'(x_n)}$$

where p is a zero of f of multiplicity m. This scheme yields faster convergence to p. On the other hand, if p is unknown, its multiplicity is unknown. So it might prove difficult to know which modified version to use.

Another method used for locating function zeros is a method attributed to Steffensen [45] which gives

$$x_{n+1} = x_n - \frac{f^2(x_n)}{f(x_n + f(x_n)) - f(x_n)}$$

This method has quadratic convergence like Newton's method, but unlike Newton's method, it does not require use of $f'(x)$, which can at times be very beneficial.

The Secant method is another that does not require the use of $f'(x)$. For this method, two initial values x_1, x_2 need to be fed into the program. Then the rest of the sequence of approximates can be computed from

$$x_{n+1} = x_n - f(x_n) \left[\frac{x_n - x_{n-1}}{f(x_n) - f(x_{n-1})} \right]$$

Thus, if $x_1 = .5$ and $x_2 = .6$ for the function above, $f(x) = x^3 + x - 1$, the third term in the sequence would be,

$$x_3 = x_2 - f(x_2) \left[\frac{x_2 - x_1}{f(x_2) - f(x_1)} \right]$$

$$= .6 - (-.184) \left[\frac{.1}{-.184 + .375} \right]$$

$$= .6963351.$$

The Secant method has a very simple geometric interpretation. If you consider the two points $(x_{n-1}, f(x_{n-1}))$, $(x_n, f(x_n))$ on the curve and connect them with a straight line, this line will pass through the x-axis at a point, denoted by x_{n+1}, which is closer to p than either x_n or x_{n-1} as shown in Fig. 5-25.

The line has a slope m equal to

$$m = \frac{f(x_n) - f(x_{n-1})}{x_n - x_{n-1}}$$

Because the line intersects the x-axis at $(x_{n+1}, 0)$, this gives

$$\frac{f(x_n) - 0}{x_n - x_{n+1}} = m$$

or

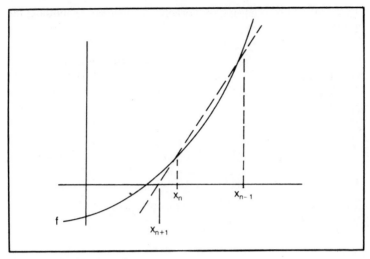

Fig. 5-25. Secant Method.

$$x_{n+1} = x_n - \frac{f(x_n)}{m}$$

$$= x_n - f(x_n) \left[\frac{x_n - x_{n-1}}{f(x_n) - f(x_{n-1})} \right]$$

In programming the Secant method, you must be careful not to perform a division by either zero or a number approximately equal to zero. This means caution should be exercised with regards to the term $f(x_n) - f(x_{n-1})$. A test should be made after each value of x_n is computed to determine whether $f(x_n)$ is close to zero. If it is close enough to zero, the program should be terminated. Otherwise $f(x_{n+1})$ will be closer to zero; the difference $f(x_{n+1}) - f(x_n)$ will be even closer yet; and the division

$$\frac{x_{n+1} - x_n}{f(x_{n+1}) - f(x_n)}$$

could produce a computer overflow.

The Secant method converges quite rapidly when conditions are right. The convergence is faster than that of the bisection method, but not as fast as the quadratic convergence of Newton or

Steffensen. In fact, the error $|x_{n+1} - p|$ is bounded above [15] by

$$|x_{n+1} - p| \leq C \, | \, x_n - p \, |^\alpha$$

where C is some positive constant, and $\alpha = (1 + \sqrt{5})/2$.

The final two methods to be mentioned in this section are methods that converge to zeros faster than Newton or Steffensen. They converge at a cubic rate, thus

$$| \, x_{n+1} - p \, | \leq C \, | \, x_n - p \, |^3.$$

The first is known as Olver's method, given by

$$x_{n+1} = x_n - \frac{f(x_n)}{f'(x_n)} - \frac{1}{2} \left[\frac{f^2(x_n)f''(x_n)}{(f'(x_n))^3} \right].$$

The other is Halley's method (named after the astronomer, Edmund Halley—see [6]), given by

$$x_{n+1} = x_n - \cfrac{f(x_n)}{f'(x_n) - \left[\cfrac{f(x_n)f''(x_n)}{2f'(x_n)} \right]}.$$

These two schemes offer such rapid convergence because of the presence of the factors f' and f''.

It might be interesting to compare some results from these methods as applied to a common problem. The first five terms in $\{x_n\}$ where tabulated in Table 5-5 for Newton's method and the bisection method with regards to $f(x) = x^3 + x - 1$ on $[0,1]$. Table 5-6 completes the comparison.

Exercises

1. The graphs of $y = x^3$ and $y = e^x$ intersect twice in quadrant 1, at points (p_1, q_1) and (p_2, q_2). Write a program that approximates p_1 and p_2 by locating the zeros to

$$f(x) = e^x - x^3.$$

Use all six methods discussed above: Newton, Bisection, Secant, Steffensen, Olver, and Halley, and compare the results.

Table 5-6. Comparing Four Approximates of the Zero to f(x).

n	Steffensen x_n	Secant x_n
1	.5	.5
2	.7823529	.6
3	.7080930	.6963351
4	.6842015	.6813178
5	.6823380	.6823158

n	Halley x_n	Olver x_n
1	.5	.5
2	.6810345	.7536443
3	.6823278	.6826707
4	.6823280	.6822058
5	.6823278	.6823278

Terminate each scheme when $|f(x_n)| < .000001$. Use the derivative formulas $f'(x) = e^x - 3x^2$ and $f''(x) = e^x - 6x$. Do the rates of convergence agree with what they are supposed to be? Be careful in choosing your starting values.

2. Find the exact roots to the quintic equation

$$x^5 - 10x^4 + 35x^3 - 50x^2 + 25x - 2 = 0.$$

NUMERICAL INTEGRATION

In analysis the need often arises to evaluate the definite integral of a function f that either has no antiderivative expressable in closed form or whose antiderivative may not be obtained by elementary or standard methods. One viable alternative to this is simply to obtain a numerical approximation to the value of the integral,

$$\int_a^b f(x)dx.$$

The standard procedure is to use a finite sum of the form

$$\sum_{i=0}^{n} a_i f(x_i).$$

To begin, assume that f is a fairly "nice" function on the interval [a,b], nice in the sense that it is at least continuous and preferably differentiable several times. The first step in the numerical scheme is to select an integral value N, which represents the number of subintervals that [a,b] is to be partitioned into. Then a set $\{x_0, x_1, x_2, \cdots, x_N\}$ of distinct nodes is chosen, where $x_0 = a$, $x_N = b$ and $x_i < x_{i+1}$ for all choices of i. Typically, the value of x_i is given by

$$x_i = x_{i-1} + \frac{b-a}{N}.$$

These nodes represent the end points of the subintervals of [a,b] as shown in Fig. 5-26.

The values $f(x_i)$ represent the function value at each node. But the numbers a_i, which differ according to the method used, are the real key to approximating the true value of the integral. Three of the more popular methods of this numerical integration (also called numerical quadrature) will be presented here. All of them fall under the classification of closed Newton-Cotes methods [13].

When f is a nice function, as assumed, the integral of f can be interpreted as denoting the algebraic area of the region R bounded by the graph of f, the horizontal axis, and the lines x = a, x = b. For the function depicted in Fig. 5-27, R would be represented by the shaded region. Any region above the axis is said to have positive area, while any region below has negative area. The algebraic area of a region is the sum of its positive area and its negative area. Consequently a region could have an algebraic area of zero (this means there is an equal amount of region above the axis as there is below), or an algebraic area that is negative (more region below than above).

The problem is to approximate the algebraic area of the region R. Our first method of quadrature makes use of rectangles to

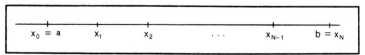

Fig. 5-26. The nodes of an interval [a,b].

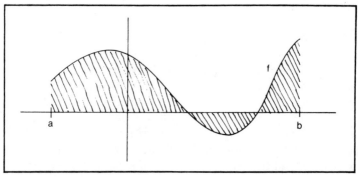

Fig. 5-27. Region of integration.

approximate the area. Assume that the nodes $\{x_0, x_1, \cdots, x_N\}$ are equally spaced, with

$$x_i - x_{i-1} = \frac{b-a}{N}$$

and these nodes are the vertices for N rectangles. Each rectangle has the same width of $x_i - x_{i-1} = (b-a)/N$, but their heights are different. The height of the first rectangle is the function value at the midpoint of the interval $[a, x_1]$; thus the height, denoted by a_1, is $a_1 = f((a + x_1)/2)$. Similarly the height, a_2, of the second rectangle is $a_2 = f((x_1 + x_2)/2)$. In general, the height of the i-th rectangle is $a_i = f((x_{i-1} + x_i)/2)$. These rectangles are depicted in Fig. 5-28.

Summing the areas of the N rectangles gives an area A_r of

$$A_r = \sum_{i=1}^{N} a_i(x_i - x_{i-1}) = \sum_{i=1}^{N} \frac{b-a}{N} \, f\left(\frac{x_{i-1} + x_i}{2}\right)$$

$$= \frac{b-a}{N} \sum_{i=1}^{N} f\left(\frac{x_{i-1} + x_i}{2}\right).$$

It's this latter expression that represents the formula, known as the Mid-Point formula, for approximating the definite integral of f,

$$A_r = \int_a^b f(x)dx.$$

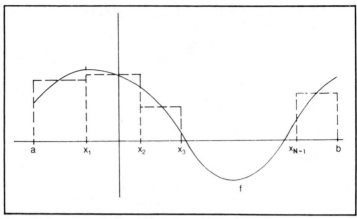

Fig. 5-28. Approximating the integral with rectangles.

You can apply this formula to the particular integral,

$$\int_0^2 (x^2 + 3)dx.$$

Choosing $N = 10$ automatically forces the nodes to be $\{0, .2, .4, \cdots, 1.8, 2.0\}$. This gives the values for the heights a_i and the areas of the 10 rectangles as listed in Table 5-7. From it you can conclude $A_r = 8.660$.

The second method of approximating integrals involves the summing of areas of trapezoids (instead of rectangles), hence it is

Table 5-7. Approximating the Integral of $f(x) = x^2 + 3$.

i	$[x_{i-1}, x_i]$	a_i	Area
1	[0, .2]	$f(.1) = 3.01$	.602
2	[.2, .4]	$f(.3) = 3.09$	.618
3	[.4, .6]	$f(.5) = 3.25$	.650
4	[.6, .8]	$f(.7) = 3.49$	.698
5	[.8, 1.0]	$f(.9) = 3.81$	.762
6	[1.0, 1.2]	$f(1.1) = 4.21$	.842
7	[1.2, 1.4]	$f(1.3) = 4.69$	.938
8	[1.4, 1.6]	$f(1.5) = 5.25$	1.050
9	[1.6, 1.8]	$f(1.7) = 5.89$	1.178
10	[1.8, 2.0]	$f(1.9) = 6.61$	1.322
			8.660

called the Trapezoidal Rule. Focus on the nodes x_{i-1} and x_i. A line segment is drawn connecting the two points $(x_{i-1}, f(x_{i-1}))$ and $(x_i, f(x_i))$. The quadrilateral produced is a trapezoid as shown in Fig. 5-29. The area of this trapezoid is an approximation for the integral,

$$\int_{x_{i-1}}^{x_i} f(x)dx.$$

The area of the trapezoid is

$$(x_i - x_{i-1}) \; \frac{f(x_{i-1}) + f(x_i)}{2}.$$

The sum of the areas of the N trapezoids gives an area A_t of

$$A_t = \sum_{i=1}^{N} (x_i - x_{i-1}) \; \frac{f(x_{i-1}) + f(x_i)}{2}$$

$$= \frac{b - a}{2N} \sum_{i=1}^{N} [f(x_{i-1}) + f(x_i)]$$

$$= \frac{b - a}{2N} [f(x_0) + 2f(x_1) + 2f(x_2) + \cdots + 2f(x_{N-1}) + f(x_N)].$$

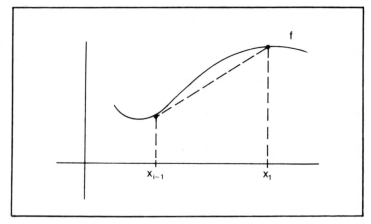

Fig. 5-29. A trapezoid.

Table 5-8. Trapezoidal Approximation to y=x²+3 on [0,2].

i	$[x_{i-1}, x_i]$	Area of i-th Trapezoid
1	[0, .2]	.2(3.02) = .604
2	[.2, .4]	.2(3.10) = .620
3	[.4, .6]	.2(3.26) = .652
4	[.6, .8]	.2(3.50) = .700
5	[.8, 1.0]	.2(3.82) = .764
6	[1.0, 1.2]	.2(4.22) = .844
7	[1.2, 1.4]	.2(4.70) = .940
8	[1.4, 1.6]	.2(5.26) = 1.052
9	[1.6, 1.8]	.2(5.90) = 1.180
10	[1.8, 2.0]	.2(6.62) = 1.324
		8.680

This is the trapezoidal approximation to the definite integral,

$$A_t = \int_a^b f(x)\ dx.$$

If you apply this formula to the integral

$$\int_0^2 (x^2 + 3)dx$$

with N = 10, you get the data listed in Table 5-8. Thus, $A_t = 8.680$, which is slightly different from $A_r = 8.660$ and in fact, is not quite as close to the exact answer of 26/3.

The third method of quadrature, called Simpson's Rule, is unique because it connects points on the curve with a parabola. First consider the three points $(x_0, f(x_0))$, $(x_1, f(x_1))$, and $(x_2, f(x_2))$. A parabola that passes through all three points as shown in Fig. 5-30 is then determined. The area of the region bounded by the parabola, the x-axis, and the lines $x = x_0$, $x = x_2$ is given by

$$\frac{b - a}{N} \left[\frac{f(x_0) + 4f(x_1) + f(x_2)}{3} \right].$$

The next step is to consider the three nodes x_2, x_3, x_4, and to pass a

255

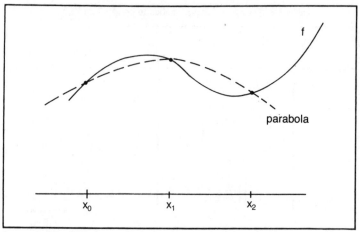

Fig. 5-30. Three points on f determine a parabola.

parabola through the three associated points on the curve as shown in Fig. 5-31. This parabola is generally different than the first one. But, again, the area of the region bounded by the parabola, the x-axis, and the lines $x = x_2$, $x = x_4$ is given by the similar formula,

$$\frac{b - a}{N} \left[\frac{f(x_2) + 4f(x_3) + f(x_4)}{3} \right].$$

Summing the two areas gives

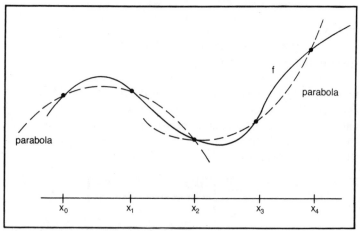

Fig. 5-31. Parabolas through 5 nodes on $y = x^2 + 3$.

256

$$\frac{b-a}{N}\left[\frac{f(x_0) + 4f(x_1) + 2f(x_2) + 4f(x_3) + f(x_4)}{3}\right]$$

The process continues with the next set of three nodes x_4, x_5, x_6, and so on until the last set of three nodes x_{N-2}, x_{N-1}, x_N. (In order to have non-overlapping parabolas you must have an even number for N, N = 2k). Then there will be k parabolas and k distinct regions, whose total area A_s is given by,

$$A_s = \frac{b-a}{3N} \sum_{i=1}^{k} [f(x_{2i-2}) + 4f(x_{2i-1}) + f(x_{2i})]$$

You may find that this latter summation formula lends itself quite well to programming. If you apply Simpson's method to the integral of $f(x) = x^2 + 3$ on [0,2] with n = 10, you get the data as listed in Table 5-9.

Thus $A_s = 8.66665$, which practically coincides (roundoff error prevents actual coinciding) with the exact value of 26/3. In this instance, this result should be expected because $f(x) = x^2 + 3$ is itself a parabola, and hence the k parabolas that are used in the approximation method all coincide with f. In general, Simpson's method for approximating a definite integral is more accurate and has greater precision than either of the other two methods [13]. Table 5-10 depicts how these values compare to the exact value for several other specific functions. A value of N = 10 is used in all cases.

Error formulas for those three methods are commonly studied in a class on calculus, or numerical analysis. This analysis will show why Simpson's rule yields such good results.

Table 5-9. Simpson's Approximation to $y = x^2 + 3$ on [0,2].

i	$[x_{2i-2}, x_{2i}]$	Area
1	[0, .4]	(1/15)(3 + 12.16 + 3.16) = 1.22133
2	[.4, .8]	= 1.34933
3	[.8, 1.2]	= 1.60533
4	[1.2, 1.6]	= 1.98933
5	[1.6, 2.0]	= 2.50133
		8.66665

Table 5-10. Comparing the Three Quadrature Methods.

	$\int_0^2 \sin(x)dx$	$\int_1^2 \ln(x)\ dx$
Exact Value of Integral	1.4161468	.3862944
A_r	1.4185098	.3865024
A_t	1.4114232	.3858780
A_s	1.4161595	.3862934

Exercises

1. Write a program which approximates the integral of a function f over an interval [a,b] by employing all three of the quadrature methods. Test your program with the specific integral,

$$\int_0^2 \sqrt{x + \sin(x)}\ dx.$$

You may wish to use a varying set of values for N.

2. Integrate $f(x) = x^2 + 3$ over the interval [0,2] by using the Mid-Point Rule. What expression would represent the value of A_r for the general case with N rectangles, evenly spaced apart? (hint: it might prove useful to know the formula,

$$1^2 + 2^2 + \cdots + k^2 = \frac{k(k+1)\ (2k+1)}{6}\ ).$$

ROMBERG INTEGRATION

The methods of quadrature discussed in the previous section, the Mid-Point, Trapezoidal, and Simpson's Rule, were discussed using equally spaced nodes. In other words, $x_1 - x_0 = x_2 - x_1 = \cdots = x_n - x_{n-1} = (b-a)/n$. Actually, there was no special reason for this property other than to help simplify the arithmetic involved. This is helpful to those programming a computer to carry out the computations. In this case, the nodes are usually included in a loop, and each passage through the loop corresponds to a common increment, say h, added to the previous node. This could appear in your program as illustrated by the following set of statements.

```
100   DIM X(100)
120   INPUT N,A,B
130   X(0) = A
140   H = (B − A)/N
150   FOR I = 1 TO N
160       X(I) = X(I−1) + H
            •
            •
            •
200   NEXT I
```

Equal spacing of the nodes causes little difficulty or problems for the multitude of functions commonly presented in a class. You should be aware though, that there are many functions that are badly "misbehaved" and whose graphs are extremely oscillatory over particular intervals. (See Fig. 5-32). In this situation, equally spaced nodes would not be suitable. Instead, you would want a greater concentration of nodes on intervals where the function oscillates most widely. But how are you to tell beforehand where these intervals are? There would have to be some testing done in the program to help determine the amount of oscillation between successive nodes. You would insert additional nodes on intervals in which the oscillation exceeds a given tolerance. This form of numerical integration is commonly listed under the title of adaptive quadrature [13].

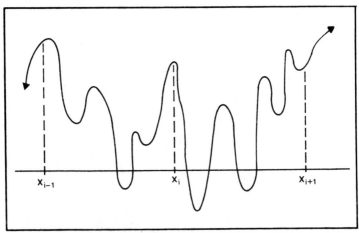

Fig. 5-32. Highly oscillatory function.

You can help circumvent this problem by relying on an improvement of the quadrature methods discussed in the previous section. One such improvement is known as Romberg integration [45]. Romberg integration can be applied to either the Mid-Point, Simpson's, or the Trapezoidal method. The remainder of this section will concentrate only on the Trapezoidal method, although the others generalize similarly.

$A_t(n)$ indicates the sum of the areas of the n trapezoids with evenly spaced nodes, $x_{i+1} - x_i = (b - a)/n$. Recall that this sum is equal to,

$$A_t(n) = \frac{(b - a)}{2n} \left[f(x_0) + 2 \sum_{i=1}^{n-1} f(x_i) + f(x_n) \right].$$

The underlying principle here is that the difference between the true value, A, of the integral

$$\int_a^b f(x)dx$$

and this expression $A_t(n)$, can be expressed as a series in powers of $(1/n)^2$; thus,

$$A - A_t(n) = \frac{a_1}{n^2} + \frac{a_2}{n^4} + \frac{a_3}{n^6} + \cdots$$

for the appropriate choices of a_i.

In the method of Romberg integration, the idea is to eliminate the successive terms a_1/n^2, a_2/n^4, $\cdots$, which in turn improves the new estimate for A. Setting n equal to the values 1, 2, 4, 8, $\cdots$ gives, in particular, the formulas,

1. $A - A_t(1) = a_1 + a_2 + a_3 + \cdots$
2. $A - A_t(2) = a_1/4 + a_2/16 + a_3/64 + \cdots$
3. $A - A_t(4) = a_1/16 + a_2/256 + a_3/4096 + \cdots$
4. $A - A_t(8) = a_1/64 + a_2/2^{12} + a_3/2^{18} + \cdots$.

Multiplying equation number 2 by 4, and subtracting from number 1, will eliminate a_1, and give the equation,

$$A = \frac{4A_t(2) - A_t(1)}{3} - \frac{1}{4}a_2 - \frac{5}{16}a_3 - \cdots .$$

Likewise, eliminating a_1 from equations 2 and 3 gives,

$$A = \frac{4A_t(4) - A_t(2)}{3} - \frac{1}{64}a_2 - \frac{5}{1024}a_3 - \cdots$$

and eliminating a_1 from equations 3 and 4 gives,

$$A = \frac{4A_t(8) - A_t(4)}{3} - \frac{1}{2^{10}}a_2 - \frac{5}{2^{16}}a_3 - \cdots .$$

To simplify the notation, define $A_t(n,2n)$ to be

$$A_t(n,2n) = \frac{4A_t(2n) - A_t(n)}{3} .$$

Now, having eliminated a_1 from the picture, you can employ the same process to remove a_2 from the new set of equations, of which the first three are,

5. $A = A_t(1,2) - a_2/2^2 - 5a_3/2^4 - \cdots$
6. $A = A_t(2,4) - a_2/2^6 - 5a_3/2^{10} - \cdots$
7. $A = A_t(4,8) - a_2/2^{10} - 5a_3/2^{16} - \cdots .$

If you multiply equation 6 by 16 and then subtract from equation 5, the resulting value for A will be

$$A = \frac{16A_t(2,4) - A_t(1,2)}{15} + \frac{1}{2^6}a_3 + \cdots .$$

Similarly, multiplying equation number 7 by 16, and subtracting from number 6, yields

$$A = \frac{16A_t(4,8) - A_t(2,4)}{15} + \frac{1}{2^{12}}a_3 + \cdots .$$

Define $A_t(n,2n,4n)$ to be

$$A_t(n,2n,4n) = \frac{16A_t(2n,4n) - A_t(n,2n)}{15} .$$

Proceeding further you would get,

$$A_t(n,2n,4n,8n) = \frac{64A_t(2n,4n,8n) - A_t(n,2n,4n)}{63}$$

which results from eliminating a_3 from the set of equations. At this stage in the process, you are able to express the value A as shown in the following series of equations,

$$A = A_t(1,2,4,8) - \frac{1}{2^{12}} a_4 - \bullet\bullet\bullet$$

$$A = A_t(2,4,8,16) - \frac{1}{2^{20}} a_4 - \bullet\bullet\bullet$$
$$\bullet$$
$$\bullet$$
$$\bullet$$

These arrays serve as approximates for the integral A. In general, these approximates have the form

$$A_t(n,2n,2^2n,\bullet\bullet\bullet,2^m n) = \frac{2^{2m}A_t(2n,2^2n,\bullet\bullet\bullet,2^m n) - A_t(n,2n,2^2n,\bullet\bullet\bullet,2^{m-1}n)}{2^{2m} - 1}$$

These approximates are often presented in a table of the form shown in Table 5-11. You then investigate whether the entries in

Table 5-11. An Array of Romberg Approximates.

$A_t(1)$				
$A_t(2)$	$A_t(1,2)$			
$A_t(4)$	$A_t(2,4)$	$A_t(1,2,4)$		
$A_t(8)$	$A_t(4,8)$	$A_t(2,4,8)$	$A_t(1,2,4,8)$	
$A_t(16)$	$A_t(8,16)$	$A_t(4,8,16)$	$A_t(2,4,8,16)$	$A_t(1,2,4,8,16)$

any row tend to a limit (when adjacent values are close), and if so, this limit is the value of the integral,

$$A = \int_{a}^{b} f(x)dx.$$

This method can be demonstrated for the function $f(x) = x^2 + 3$ on the interval $[0,2]$, as discussed in the previous section. First compute the values $A_t(1)$, $A_t(2)$, $A_t(4)$ and $A_t(8)$. For $n = 1$, you will get

$$A_t(1) = \frac{2}{2}[f(0) + f(2)] = 10.$$

For $n = 2$ you will get

$$A_t(2) = \frac{1}{2}[f(0) + 2f(1) + f(2)] = 9.$$

For $n = 4$ you will get

$$A_t(4) = \frac{1/2}{2}[f(0) + 2f(1/2) + 2f(1) + 2f(3/2) + f(2)]$$

$$= 35/4.$$

For $n = 8$ you will get

$$A_t(8) = 139/16.$$

Then the values of the 2-tuples are

$$A_t(1,2) = \frac{4A_t(2) - A_t(1)}{3} = 26/3$$

$$A_t(2,4) = \frac{4A_t(4) - A_t(2)}{3} = 26/3$$

$$A_t(4,8) = \frac{4A_t(8) - A_t(4)}{3} = 26/3.$$

The values of the 3-tuples are

$$A_t(1,2,4) = \frac{16A_t(2,4) - A_t(1,2)}{15} = 26/3$$

$$A_t(2,4,8) = \frac{16A_t(4,8) - A_t(2,4)}{15} = 26/3.$$

Furthermore, $A_t(1,2,4,8) = [64(26/3) - 26/3]/63 = 26/3$. Arranging these scores in tabular form gives Table 5-12, where in the third row, there are two consecutive scores that are the same; hence this value is the value for A. Note that convergence came quickly for this example, because of the elementary nature of the function on the interval.

10			
9	26/3		
35/4	26/3	26/3	
139/16	26/3	26/3	26/3

Table 5-12. Romberg Approximates for $f(x) = x^2 + 3$.

In programming this method, you would initially build in some tolerance level, $\epsilon > 0$, so that when two consecutive row scores from the table differ by less than ϵ, the right-most score would be used as the integral approximate. Further computations would not need to be carried out. To do the computations, you would first set n = 1, and compute $A_t(1)$. Then, set n = 2 and compute $A_t(2)$, and $A_t(1,2)$. If the difference between these two is less than ϵ, print out $A_t(1,2)$ as the answer. Otherwise set n = 4 and compute $A_t(4)$ and $A_t(2,4)$. Check the difference, and either print out $A_t(2,4)$ as the answer, or compute $A_t(1,2,4)$, and compare it with $A_t(2,4)$. If necessary, continue on with n = 8, and make successive computations and difference checks.

Unfortunately, some versions of BASIC do not allow for the arrays $A_t(1,2,4)$, or $A_t(2,4,8)$ or $A_t(1,2,4,8)$ because the variable A_t (in programming this could be denoted by AT, or AREAT) has more than two subscripts appended to it. You will have to think up some way to get around this obstacle. One possible way is to replace A_t with the 2-dimensional array M(i,j) where, for instance,

264

$$A_t(1) = M(1,1) \qquad A_t(1,2) = M(2,1) \qquad A_t(1,2,4) = M(3,1)$$
$$A_t(2) = M(1,2) \qquad A_t(2,4) = M(2,2) \qquad A_t(2,4,8) = M(3,2)$$
$$A_t(4) = M(1,4) \qquad A_t(4,8) = M(2,4)$$
$$A_t(8) = M(1,8) \qquad A_t(8,16) = M(2,8)$$

while in general,

$$M(i,j) = A_t(j, 2^1j, 2^2j, \cdots, 2^{i-1}j).$$

The first subscript i will indicate how many subscripts A_t has, while j (which must be a power of 2) denotes the value of the first subscript of A_t.

Exercises

1. Write a program which approximates the integral

$$\int_a^b f(x)dx$$

by Romberg integration. Test your program for the function

$$f(x) = 3x\sin(2x)$$

on the interval $[0, \pi/2]$. Use $\epsilon = .001$.

2. Apply Romberg integration to refine Simpson's method. Since the error in Simpson's method is proportional to n^{-4}, this allows you to write

$$A - A_s(n) = a_1/n^4 + a_2/n^6 + \cdots.$$

If you proceed as with the trapezoidal method, what is the expression that $A_s(n, 2n, 4n)$ represents?

THE BETA FUNCTION

One of the most useful functions studied in applied science is that of the gamma function, $\Gamma(x)$. This renowned function, defined by

$$\Gamma(x) = \int_0^\infty e^{-t}t^{x-1}dt$$

is attributed to Euler and is referred to as Euler's second integral [3]. Euler was searching for a function that would extend the factorial function, f(n) = n!, to a continuum of real numbers. The gamma function supplied the generalization because of the relationship $\Gamma(x+1) = x\,\Gamma(x)$. Thus, for x = 5, you have

$$\begin{aligned}
(6)\ \Gamma &= 5\,\Gamma(5) \\
&= 5{\cdot}4\,\Gamma(4) \\
&= 5{\cdot}4{\cdot}3\,\Gamma(3) \\
&= 5{\cdot}4{\cdot}3{\cdot}2\,\Gamma(2) \\
&= 5{\cdot}4{\cdot}3{\cdot}2{\cdot}1\,\Gamma(1)
\end{aligned}$$

and because $\Gamma(1) = 1$, you have $\Gamma(6) = 5!$. In general, for integral n, you have

$$\Gamma(n+1) = n!.$$

The gamma function duly serves those areas of science that are involved with evaluating improper integrals. Several examples include,

$$\int_0^1 \frac{dx}{(1-x^4)^{1/2}} = \frac{1}{4}\,\frac{\Gamma(1/4)\,\Gamma(1/2)}{\Gamma(3/4)}$$

$$\int_0^{\pi/2} \sqrt{\sin(\theta)d\theta} = \frac{\Gamma(3/4)\,\Gamma(1/2)}{2\Gamma(5/4)}\ .$$

Some of the probability distribution functions in statistics involve the gamma function. The F distribution and chi-square distribution are two such examples. The chi-square distribution is often encountered in large sample theory as an asymptotic distribution of a statistic (such as Pearson's test for goodness-of-fit) used in making inferences. The density function of the chi-square distribution with k degrees of freedom is given by

$$f(x) = \frac{x^{(k/2)-1}e^{-x/2}}{2^{k/2}\,\Gamma(k/2)}\ , x > 0.$$

Even-order moments of a normal distribution can also be expressed in terms of gamma functions. For, if X has the standard normal distribution, the 2k-th moment of X is,

$$E(X^{2k}) = \frac{1}{\sqrt{2\pi}} \int_{-\infty}^{\infty} x^{2k} e^{-x^2/2} \, dx$$

$$= \sqrt{(2/\pi)} \int_{0}^{\infty} x^{2k} e^{-x^2/2} dx$$

$$= \frac{2^k}{\sqrt{\pi}} \Gamma(k + 1/2).$$

In particular, with $k = 0$,

$$1 = E(X^0) = \frac{1}{\sqrt{\pi}} \Gamma(1/2)$$

so $\Gamma(1/2) = \sqrt{\pi}$, which is a well known property of the gamma function.

The gamma function, as defined above, is actually meaningful for all positive $x > 0$, but the definition can be extended, by means of the equation

$$\Gamma(x+1) = x \, \Gamma(x)$$

to include certain negative values. It will be useful for the remainder of this section to refer to certain values of $\Gamma(x)$, so a rough graph of the function is shown in Fig. 5-33.

A function that is closely related to Γ is the beta function B given by

$$B(x,y) = \int_{0}^{1} t^{x-1} (1-t)^{y-1} dt.$$

This function was also studied by Euler, and in fact is known as the first Eulerian integral. When the two variables are positive, $x,y > 0$, the function is a well-behaved, continuous function. Furthermore, if

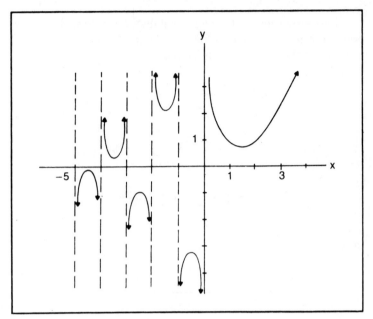

Fig. 5-33. The gamma function.

x and y are both natural numbers, say x = n, y = m, the integrand is simply a polynomial in t. Hence it can easily be integrated. For if n = 3 and m = 2 then

$$B(3,2) = \int_0^1 t^2(1-t)^1 dt$$

$$= \int_0^1 (t^2 - t^3) dt$$

$$= (1/3) - (1/4)$$

$$= 1/12.$$

It is interesting to note that if you interchange the two variables, the resulting function value is the same. Notice that

$$B(2,3) = \int_0^1 t(1-t)^2 dt$$

$$= \int_0^1 (t - 2t + t^3)dt$$

$$= (1/2) - (2/3) + (1/4)$$

$$= 1/12.$$

Consequently, one very important property of the beta function is this symmetry relation, $B(x,y) = B(y,x)$.

Another important relation involves the basic trigonometric functions, $\sin(\theta)$ and $\cos(\theta)$. Using the transformation $t = \sin^2(\theta)$, you have

$$B(x,y) = \int_0^1 t^{x-1}(1-t)^{y-1}dt$$

$$= \int_0^{\pi/2} \sin^{2x-2}(\theta)\cos^{2y-2}(\theta)2\sin(\theta)\cos(\theta)d\theta$$

$$= 2\int_0^{\pi/2} \sin^{2x-1}(\theta)\cos^{2y-1}(\theta)d\theta.$$

As a consequence of this, setting $y = 1/2$ and $x = (n+1)/2$, you have

$$\int_0^{\pi/2} \sin^n(\theta)d\theta = \frac{1}{2} B\left(\frac{n+1}{2}, \frac{1}{2}\right)$$

The discussion on B can be aided by graphing. The curve $z = B(x,y)$, which is actually a surface in 3-space, is a nice smooth surface throughout the entire first quadrant as shown in Fig. 5-34. The surface is asymptotic to the plane $z = 0$. This follows because on the planes $y = 1, 2, 3, \cdots$ you have

$$\lim_{x \to \infty} B(x,1) = \lim_{x \to \infty} \frac{1}{x} = 0$$

$$\lim_{x \to \infty} B(x,2) = \lim_{x \to \infty} \frac{1}{x(x+1)} = 0$$

$$\vdots$$

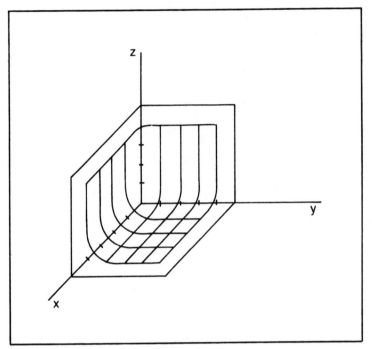

Fig. 5-34. The beta function in quadrant 1.

$$\lim_{x \to \infty} B(x,n) = \lim_{x \to \infty} \frac{1}{x(x+1) \cdots (x+n-1)} = 0.$$

The surface is also asymptotic to the other two planes $x = 0$ and $y = 0$. And because $B(x,y) = B(y,x)$, the surface is symmetric with respect to the plane $y = x$.

The most important property involving B is its relationship to the gamma function. This characteristic is summarized as

$$B(x,y) = \frac{\Gamma(x) \, \Gamma(y)}{\Gamma(x+y)}$$

from which you could readily verify the earlier result

$$B(3,2) = \frac{\Gamma(3) \, \Gamma(2)}{\Gamma(5)}$$

$$= \frac{2!\ 1!}{4!}$$
$$= 1/12.$$

Proofs of this interesting relationship (now you know why Euler studied the Beta function) are to be found in almost all texts on advanced calculus, such as [27], where the usual proof depends on evaluating a double iterated integral. A different proof appears in [75] and requires Laplace transforms.

The above relation is valid for all positive x and y, but it also serves as the most commonly used means for extending the domain of B to include negative values. B(−1/2,−1/3) is defined to equal

$$B\left(\frac{-1}{2}, \frac{-1}{3}\right) = \frac{\Gamma\left(\frac{-1}{2}\right)\Gamma\left(\frac{-1}{3}\right)}{\Gamma\left(\frac{-5}{6}\right)}.$$

Allowing x,y to assume negative values causes a tremendous change in the behavior of B(x,y). This is clearly due to the discontinuous action of Γ for negative values. The surface of z = B(x,y) will oscillate wildly between $+\infty$ and $-\infty$ for negative x,y and will be undefined when x and y are negative integers. Furthermore, for those values of x,y where x + y is a negative integer but neither x nor y is, then Γ (x) and Γ (y) will be infinite values, but Γ (x + y) will equal $\pm\infty$: hence B(x,y) = 0. These are the points where the surface intersects the plane z = 0. Sample points include (−4/3,1/3), or (−.4,−.6). These points belong to the family of lines x + y = −n for n = 0,1,2,•••, and are represented as in Fig. 5-35. The points that are circled represent the integral lattice points and are the points where B is undefined, since Γ (x) Γ (y) = $\pm\infty$ and Γ (x + y) = $\pm\infty$. The points (x,y) where either x or y is a negative integer or zero but not both, imply that one of Γ (x) or Γ (y) is infinite, but Γ (x + y) is finite. Hence B(x,y) is $\pm\infty$ (the algebraic sign is determined by the manner one approaches (x,y)). The totality of these points that cause the surface z = B(x,y) to extend infinitely far is represented by the darkened stripes of horizontal and vertical lines, as shown in Fig. 5-36.

The bizarre behavior of B can be brought to light by inspecting

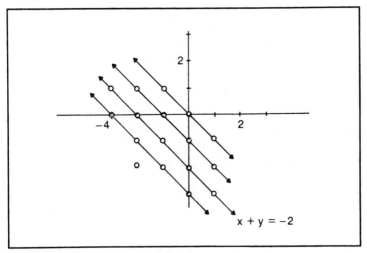

Fig. 5-35. Some zeros of the beta function.

one of the unit regions. An enlarged unit is shown in Fig. 5-37. As the point (x,y) approaches a horizontal or vertical line x = n or y = m, the values of z tend to ± ∞, but as (x,y) approaches the diagonal line, z tends to zero. So what happens to B(x,y) as (x,y) approaches

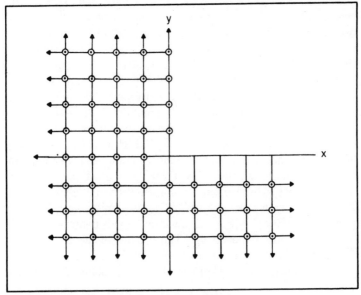

Fig. 5-36. Points where B(x,y) extends to infinity.

272

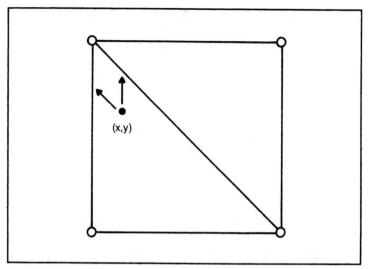

Fig. 5-37. A unit region.

the corner point? More detailed discussion of the beta function can be found in [59].

Let us return to the case $x > 0$ and $y > 0$ and compute a few values of $B(x,y)$. First, if you set $y = 2$, and let $x = 1,2,3,\cdots$ you get

$$
\begin{aligned}
B(1,2) &= 1/2 \\
B(2,2) &= 1/6 \\
B(3,2) &= 1/12 \\
B(4,2) &= 1/20 \\
B(5,2) &= 1/30 \\
B(6,2) &= 1/42
\end{aligned}
$$

where, in general, $B(n,2) = 1/(n(n+1))$. This means that the ratio of successive values, $B(n+1,2)$ to $B(n,2)$, can be simplified to

$$
\frac{B(n+1,2)}{B(n,2)} = \frac{n(n+1)}{(n+1)(n+2)} = \frac{n}{n+2}
$$

or, equivalently,

$$
B(n+1,2) = \frac{n}{n+2} \, B(n,2).
$$

If you replace $y = 2$ with $y = 3$, and repeat the procedure, you get

$$
\begin{aligned}
B(1,3) &= 1/3 = 2/6 \\
B(2,3) &= 1/12 = 2/24 \\
B(3,3) &= 1/30 = 2/60
\end{aligned}
$$

$$\vdots$$

$$B(n,3) = \frac{2}{n(n+1)(n+2)}$$

so,

$$B(n+1,3) = \frac{n}{n+3} \; B(n,3).$$

This pattern is beginning to take the form,

$$B(n+1,m) = \frac{n}{n+m} \; B(n,m)$$

for integral values of n and m. You might be interested in knowing whether this relationship would hold for any positive real variables, i.e.,

$$B(x+1,y) = \frac{x}{x+y} \; B(x,y).$$

You can gather data on this by writing a program that will approximate $B(x+1,y)$ and $B(x,y)$, and compare their ratio with $x/(x+y)$. In general, since $B(x,y)$ is defined as a definite integral over $[0,1]$, the integral can be approximated by the previously discussed Simpson's method. Consequently, dividing $[0,1]$ into n (an even number) equal parts,

$$0 = t_0 < t_1 < t_2 < \cdots < t_n = 1,$$

you have

$$B(x,y) = \int_0^1 t^{x-1}(1-t)^{y-1}dt$$

$$= \frac{1}{3} \left[f(t_0) + 4f(t_1) + 2f(t_2) + 4f(t_3) + \cdots + f(t_n) \right] \frac{1}{n}$$

where $f(t) = t^{x-1}(1-t)^{y-1}$. You will need to input x,y, and probably a tolerance figure $\epsilon > 0$ so that the approximated values for n and n+2 partitions are within ϵ of one another.

EXERCISES

1. Write a program which approximates B(x,y) and B(x+1,y) by using Simpson's Rule. Run the program using values $\epsilon = .001$, x = 3.5, and y = 2.7. Does it seem to follow that

$$B(4.5,2.7) = \frac{3.5}{6.2} B(3.5,2.7)?$$

2. Compute the two partial derivatives $\frac{\partial}{\partial x} B(x,y)$ and $\frac{\partial}{\partial y} B(x,y)$.

THE ZETA FUNCTION

In elementary courses in calculus, the student is usually exposed to some standard infinite series that can be used to aptly illustrate key notions on convergence. One such series is the geometric series of ratio 1/2,

$$\sum_{n=0}^{\infty} (1/2)^n .$$

Another is the harmonic series,

$$\sum_{n=1}^{\infty} 1/n.$$

The harmonic series, which is the classic series for a divergent series whose terms are positive and decrease to zero, is actually a special case (s = 1) for the hyperharmonic series,

$$\sum_{n=1}^{\infty} (1/n)^s.$$

The ratio test fails for this series but, applying the integral test, you get

$$\int_1^\infty \frac{1}{x^s}\,dx = \lim_{M\,\infty} \left.\frac{x^{1-s}}{1-s}\right|_1^M = \lim_{M\to\infty} \frac{M^{1-s}}{1-s} - \frac{1}{1-s}$$

$$= \left[\lim_{M\to\infty} \frac{M^{1-s}}{1-s}\right] - \frac{1}{1-s}.$$

From this you can see that the integral is finite and equals $1/(1-s)$ if $s > 1$, and is infinite if $s \le 1$. Consequently, the hyperharmonic series is convergent only for $s > 1$. The series, when viewed as a function of the exponent s, is what has come to be known as the zeta function ζ (s), so named by Riemann:

$$\zeta(s) = \sum_{n=1}^\infty (1/n)^s.$$

An extensive historical background provides ample literature on the zeta function [4]. Noteworthy here is the work done on determining the zeros of the function, its relationship with other transcendental functions like $\sin(x)$, e^x, and $\mathrm{gamma}(x)$, and the closed expressions for particular values of the function.

Within this final topic, it's the particular value $\zeta(2)$ that had been most sought after. Mathematicians had been trying for a number of years to determine the explicit value of the series

$$\sum_{n=1}^\infty 1/n^2.$$

The problem was posed as early as 1650 in a book *Novae Quadraturae Arithmeticae* by the Italian, Pietro Mengoli. Many of the famous European mathematicians worked on the problem, including such notables as Wallis, Leibniz, Bernoulli, and Goldbach. Daniel Bernoulli apparently had a method for quickly computing an approximation to $\zeta(2)$, and gave as a rough approximate the value 8/5. Christian Goldbach was able to establish the inequality

$$\frac{41}{25} < \zeta(2) < \frac{5}{3}$$

but it was left to the great Swiss mathematician Euler to furnish the world with the exact value, namely $\pi^2/6$. At first, the best Euler could do was to approximate $\zeta(2)$ by the decimal,

$$\zeta(2) = 1.64493406684822643647.$$

But then, using the function $f(x) = 1 - \sin(x)/\sin(\alpha)$, where α is some constant other than a multiple of π, and using the series expansion for $\sin(x)$, Euler was able to express f as the infinite product

$$f(x) = \prod_{k=1}^{\infty}\left(1 - \frac{x}{a_k}\right)$$

where the roots of f are $\{2n\pi + \alpha, 2n\pi + \pi - \alpha\}$ for $n = 0, \pm 1, \pm 2,$ $\cdots$. Euler was then able to derive the formula

$$\frac{1}{\sin^2(\alpha)} = \frac{1}{\alpha^2} + \sum_{n=1}^{\infty}\left[\frac{1}{((2n-1)\pi-\alpha)^2} + \frac{1}{((2n-1)\pi+\alpha)^2} + \right.$$
$$\left. \frac{1}{(2n\pi+\alpha)^2} + \frac{1}{(2n\pi-\alpha)^2}\right].$$

And what a key formula this is! If you set $\alpha = \pi/2$, you get

$$\frac{1}{1} = \frac{4}{\pi^2} + \left[\frac{4}{\pi^2} + \frac{4}{9\pi^2} + \frac{4}{25\pi^2} + \frac{4}{9\pi^2}\right] + \left[\frac{4}{25\pi^2} + \frac{4}{49\pi^2} + \right.$$
$$\left. \frac{4}{81\pi^2} + \frac{4}{49\pi^2}\right] + \cdots$$

$$= \frac{4}{\pi^2}\left[2 + \frac{2}{9} + \frac{2}{25} + \frac{2}{49} + \cdots\right]$$

$$= \frac{8}{\pi^2}\left[1 + \frac{1}{3^2} + \frac{1}{5^2} + \frac{1}{7^2} + \cdots\right]$$

$$= \frac{8}{\pi^2}\left[\zeta(2) - \frac{1}{4}\zeta(2)\right]$$

$$= \frac{6}{\pi^2} \; \zeta(2)$$

from which $\zeta(2) = \pi^2/6$.

Similarly, Euler was able to verify $\zeta(4) = \pi^4/90$; he evaluated $\zeta(6)$ and $\zeta(8)$ and said that in general, for any positive integer n, $\zeta(2n) = N \cdot \pi^{2n}$ where N is some rational number. This is quite a remarkable result for ζ evaluated at any even number. You may wish to refer to the references [2,61] for further explanations on this subject.

One of the classic results from mathematics asserts that the value of zeta at an even number is related to a set of numbers B_i, called the Bernoulli numbers. These numbers are formally defined as the coefficients in the relation

$$\frac{x}{e^x-1} = B_0 + B_1 x + \frac{B_2 x^2}{2!} + \frac{B_3 x^3}{3!} + \cdots \quad |x| < 2\pi$$

with the first few values of B_i being listed in Table 5-13. The formula, attributed again to Euler, that relates the Bernoulli numbers to values of the zeta function is given by [7] the following,

$$\zeta(2n) = \frac{(-1)^{n-1}(2\pi)^{2n}B_{2n}}{2(2n)!} \; .$$

Thus, for $n = 1$, $\zeta(2) = (-1)^0(2\pi)^2 B_2/4 = \pi^2/6$, and for $n = 2$, $\zeta(4) = \pi^4/90$.

What about $\zeta(3)$? Or $\zeta(5)$? Euler was unable to determine the exact value of $\zeta(2n+1)$ for $n = 1,2,3,\cdots$. This still remains a major

i	B_i
0	1
1	$-1/2$
2	$1/6$
3	0
4	$-1/30$
5	0
6	$1/42$
7	0

Table 5-13. Bernoulli Numbers.

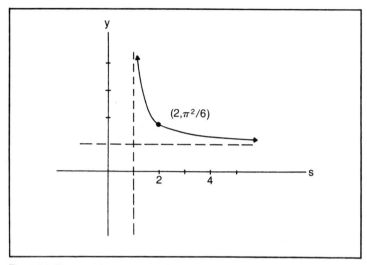

Fig. 5-38. The zeta function ζ (s), for s > 1.

unsolved problem today. The conjecture here is that ζ (2n+1) is of the form

$$\zeta(2n+1) = f(\ln(2)) \cdot \pi^{2n+1}.$$

In other words, zeta of an odd integer is equal to a product in which one of the factors is π raised to an exponent equal to the odd integer, and the other factor is some function of the natural logarithm of 2. For instance, $\zeta(3)$ is approximately equal to 1.202, and it could be that $\zeta(3)$ is related to $\ln(2)$ and π^3 by the following formula,

$$\zeta(3) = \frac{5}{43} \ (\ln(2))^3 \pi^3.$$

Likewise, $\zeta(5)$ is supposedly equal to some function of $\ln(2)$, times π^5. You may wish to compile data on the approximate values of $\zeta(3)$, $\zeta(5)$, $\zeta(7)$, and so on, and try to discover some appropriate closed expression for these particular values. As it stands now, the graph of $\zeta(s)$, for s > 1, can be sketched, as shown in Fig. 5-38. The curve has a vertical asymptote at x = 1 and a horizontal asymptote at y = 1.

Euler was able to express $\zeta(s)$ in terms of some other functions, which at times can prove to be most useful. First, he established his famous product decomposition formula, which reads

$$\zeta(s) = \frac{2^s \; 3^s \; 5^s \; 7^s \; 11^s \; \bullet\bullet\bullet}{(2^s-1)(3^s-1)(5^s-1)(7^s-1)(11^s-1)\bullet\bullet\bullet}.$$

Then he showed that

$$\zeta(s) = \frac{\theta(s)}{1 - (1/2)^s}$$

where θ is the series function

$$\theta(s) = \sum_{n=0}^{\infty} \frac{1}{(2n+1)^s}$$

This function also converges for $s > 1$, but it converges faster than ζ. It is sometimes used to obtain faster approximates for the values of ζ. For instance, to estimate $\zeta(4)$, the first 10 terms of the defining series for ζ gives

$$\sum_{n=1}^{10} \frac{1}{n^4} = 1.0820366$$

while the first 10 terms of $\theta(4)/(1 - .5^4)$ gives

$$\frac{\displaystyle\sum_{n=0}^{9} \frac{1}{(2n+1)^4}}{1 - .5^4} = \frac{1.0146573}{15/16} = 1.0823011$$

which is closer to the exact value $\pi^4/90 = 1.0823196$.

Similarly, defining the function $\emptyset$ by the alternating series

$$\emptyset(s) = \sum_{n=1}^{\infty} \frac{(-1)^{n+1}}{n^s}$$

Euler showed that ζ and $\emptyset$ are related by

$$\zeta(s) = \frac{\emptyset(s)}{1 - (1/2)^{s-1}}$$

which, even though it is similar to the above expression involving θ, has the advantage of converging for all positive $s > 0$. Using this you could supply meaning to, say, $\zeta(1/2)$ by defining

$$\zeta(1/2) = \frac{\emptyset(1/2)}{1 - 2^{1/2}}.$$

The approximate value of $\zeta(1/2)$ would be $.603/(-.414) = -1.456$.

In the middle of the nineteenth century, Riemann devoted considerable study to the zeta function. In 1859 he was able to prove a most remarkable theorem that connected the zeta function, the gamma function, and the cosine function. This equation reads,

$$\zeta(1-s) = \pi^{-s}2^{1-s} \, \Gamma(s)\cos(\pi s/2) \, \zeta(s)$$

and, in fact, this allows you to extend the definition of ζ to negative values. For instance, if $s = 2$, then

$$\zeta(-1) \quad = \pi^{-2}2^{-1} \, \Gamma(2)\cos(\pi) \, \zeta(2)$$

$$= \frac{1(-1)(\pi^2/6)}{2 \, \pi^2}$$

$$= \quad - 1/12.$$

This equation is customarily used to define ζ for all $s \neq 0$. Furthermore, in this case, since $\cos(\pi s/2) = 0$ when $s = 3,5,7,\cdots$, then $\zeta(1-s) = 0$ at these values. Thus, some of the zeros for ζ are at $1-s = -2,-4,-6,\cdots$. These are the so-called trivial zeros of the zeta function [53].

It is often the case that zeta is extended to cover complex numbers z, and hence you have

$$\zeta(z) = \sum_{n=1}^{\infty} \frac{1}{n^z}.$$

In this way zeta obeys a relationship similar to the one above, namely

$$\zeta(1-z) = \pi^{-z}2^{1-z}\ \Gamma(z)\cos(\pi z/2)\ \zeta(z).$$

This equation is sometimes useful in studying properties of the zeta function. One such property is that ζ can be expressed as the integral

$$\zeta(z) = \frac{1}{\Gamma(z)} \int_0^\infty \frac{t^{z-1}}{e^t+1}\ dt \quad \text{for Re}(z) > 0.$$

Another is that the only singularity of ζ (z) is a simple pole at $z = 1$ having residue 1. The still unproved Riemann conjecture further asserts that every non-trivial zero of ζ has a real part equal to 1/2. These statements are all of very deep mathematical origin and form the basis for current "heavy" research.

Exercises

1. Try to extend the graph of $y = \zeta(s)$ for the interval $0 < s < 1$. To this end, resort to the equation

$$\zeta(s) = \frac{\emptyset(s)}{1 - (1/2)^{s-1}}$$

which converges for $s > 0$. Write a program that approximates values of $\zeta(s)$ when $s = .1, .2, .3, \cdots, .9$. Then sketch in the graph. What happens as s tends to 1? to zero?

2. Make use of the formula

$$\frac{1}{\sin(\alpha)} = \frac{1}{\alpha} + \sum_{n=1}^\infty \left[\frac{1}{(2n-1)\pi-\alpha} - \frac{1}{(2n-1)\pi+\alpha} + \frac{1}{2n\pi+\alpha} - \frac{1}{2n\pi-\alpha} \right]$$

to compute the exact value of $1 - (1/3) + (1/5) - (1/7) + (1/9) - \cdots$.

LAGRANGE MULTIPLIERS

Let us first consider a common problem from calculus:
Find the point on the curve f, where f is given by

$$f(x) = \frac{3}{x} + \frac{x}{6}, \quad x > 0$$

that is closest to the origin.

The graph of the function f in Fig. 5-39 clearly shows that there is a definite point (a,b) that is closest to the origin. This implies that the distance between (a,b) and (0,0) is smaller than all the other distances from (x,y) to (0,0), where (x,y) lies on the graph of f. Thus, if we let g denote the (distance) function

$$g(x,y) = \sqrt{(x-0)^2 + (y-0)^2}$$
$$= \sqrt{x^2 + y^2}$$

the above calculus problem can be restated as, "find the point on the curve f that minimizes g(x,y)." You can estimate the solution (a,b) by examining the values listed in Table 5-14. Here it would appear that (a,b) lies quite close to (3/2,9/4). In fact, if you sketch the graphs of some of the level curves of g, i.e., the graphs of g(x,y) = c for, say, the five values of c as listed in Table 5-14, you would notice

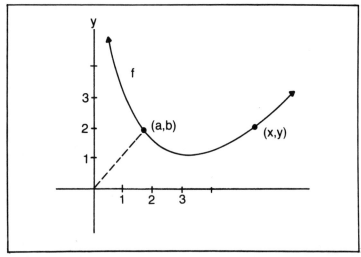

Fig. 5-39. Graph of f(x) = 3/x + x/6.

283

(x,y)	g(x,y)
(1/2, 73/12)	6.1038
(3/4, 33/8)	4.1926
(1, 19/6)	3.3208
(3/2, 9/4)	2.7042
(2, 11/6)	2.7131

Table 5-14. Values of g for f(x) = 3/x + x/6.

that the curves $g(x,y) = 2.7042$ and $f(x) = 3/x + x/6$ are practically tangent to each other as shown in Fig. 5-40 at (3/2,9/4). If (a,b) denotes the exact solution to the problem, then the curve $g(x,y) = \sqrt{a^2 + b^2}$ is tangent to f at (a,b). It is precisely this characteristic of tangency that leads into the concept of LaGrange multipliers.

The technique of LaGrange multipliers (so named after the French mathematician J. LaGrange, 1736-1813) is employed for finding the maximum or minimum value of a function g(x,y) of two variables that is defined on a curve f(x,y) = c. The important feature of this technique centers around the gradient vector of a function. This vector is always perpendicular to the curve at each point in question. In general, for a function h(x,y) of two variables, the gradient vector $\overrightarrow{\text{grad}}\ h(x,y)$ is given by

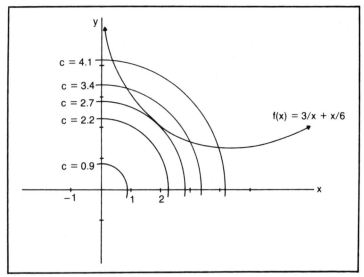

Fig. 5-40. The curves f(x) = 3/x + x/6 and g(x,y) = c.

284

$$\overrightarrow{\text{grad}} \ h(x,y) = \frac{\partial h}{\partial x}(x,y) \ i + \frac{\partial h}{\partial y}(x,y) \ j$$

and is represented pictorially in Fig. 5-41, where you can see that the gradient vector is perpendicular to the level curve of h(x,y) through (a,b). The magnitude of the gradient vector is

$$|\overrightarrow{\text{grad}} \ h(x,y)| = \sqrt{\left[\frac{\partial h}{\partial x}(x,y)\right]^2 + \left[\frac{\partial h}{\partial y}(x,y)\right]^2}.$$

To illustrate, if $h(x,y) = x^2 y$, the gradient vector at (1,3) would be given by

$$\overrightarrow{\text{grad}} \ h(1,3) = \frac{\partial h}{\partial x}(1,3)i + \frac{\partial h}{\partial y}(1,3)j$$

$$= 2xy|(1,3)_{i+x^2}|(1,3)_j$$

$$= 6i + j.$$

Since $x = 1$ and $y = 3$, the level curve of h is $h(x,y) = 1^2 \cdot 3 = 3$, which is the same curve as $y = 3/x^2$. Thus, $\overrightarrow{\text{grad}} \ h(1,3)$ is perpendicular to this curve at (1,3), has a direction $\tan^{-1}(1/6)$, and a magnitude of $\sqrt{37}$.

If two curves happen to be tangent at (a,b), their gradient vectors at that point coincide in direction although they probably

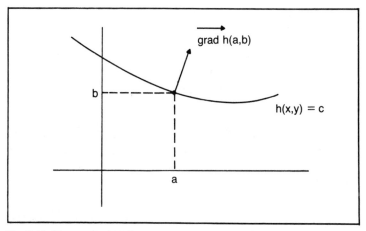

Fig. 5-41. The gradient vector.

have different magnitudes. It is the LaGrange multiplier, usually denoted by λ, that represents the ratio of these two vectors. In other words,

$$\overrightarrow{\text{grad}}\; g(x,y) = \lambda \cdot \overrightarrow{\text{grad}}\; f(x,y).$$

It follows that λ also represents the ratio of the magnitudes of the vectors.

Suppose that you wish to construct a square-base box of maximum volume. The function $g(x,y) = x^2y$ would represent the volume of the box. Furthermore, suppose that the two variables x,y satisfy the side condition $x + 2y = 5$. This equation is one of the level curves of $f(x,y) = x + 2y$, namely $f(x,y) = 5$. Since you wish to maximize $g(x,y)$ subject to $f(x,y) = 5$, the technique of LaGrange multipliers assert the existence of a real number λ for which $\overrightarrow{\text{grad}}$ $g(x,y) = \lambda \cdot \overrightarrow{\text{grad}}\; f(x,y)$. You have $\overrightarrow{\text{grad}}\; g(x,y) = 2xyi + x^2j$ and $\overrightarrow{\text{grad}}$ $f(x,y) = i + 2j$; thus the two components of the vectors must satisfy

$$2xy = \lambda \;\; (1)$$
$$x^2 = \lambda \;\; (2).$$

Solving for λ from the second equation and substituting it in the first gives $4xy = x^2$, from which either $x = 0$ or $x = 4y$. The value $x = 0$ clearly does not maximize $g(x,y)$, so x must equal $4y$. From the side condition,

$$5 = x + 2y = 47 + 2y = 6y$$

You get $y = 5/6$ and then $x = 10/3$. The maximum volume for the box is $500/54$ because $g(10/3,5/6) = 500/54$. The value of the LaGrange multiplier is $\lambda = 50/9$, which means that of the two gradient vectors, the longer one $(\overrightarrow{\text{grad}}\; g(x,y))$ is $50/9$ times as long as the short one $(\overrightarrow{\text{grad}}\; f(x,y))$. This is illustrated in Fig. 5-42.

You should realize several points about solving these max-min problems using the method of LaGrange multipliers. First, the method does extend to functions of more than two variables, such as

$$\overrightarrow{\text{grad}}\; g(x,y,z) = \lambda \cdot \overrightarrow{\text{grad}}\; f(x,y,z).$$

Secondly, it may be quite a chore, if not an impossibility, to solve the associated system of equations involving the gradient components.

286

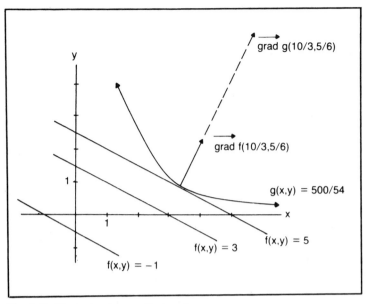

Fig. 5-42. Comparing two gradient vectors.

But this is the situation in which a scientist or an engineer might employ a computer to do some number crunching to search for an approximate solution. For instance, with the previous example of x + 2y = 5 and $g(x,y) = x^2y$, you could simply run the variable x through the set of values x = .01, .02, •••, 4.99, 5.00. You won't need to check x > 5 because then y < 0. At each stage set y equal to (5−x)/2, and evaluate g(x,y). You would only have to keep track of the largest g(x,y) and the point (a,b) that produced this largest g(x,y) value. Needless to say, you would obtain a closer approximate with smaller increment values for x. This procedure can be easily outlined by the following flowchart shown in Fig. 5-43.

A third point, in regard to LaGrange multipliers, is that you can not use this technique effectively unless you are able to form the gradient vectors, which in turn implies you must be able to compute partial derivatives. The solution point (a,b) can be approximated, as shown in the flowchart, but what about estimating the value of λ? This would actually involve more computation, because you would have to program approximate values for the partial derivatives

$$\frac{\partial f}{\partial x}(a,b), \quad \frac{\partial f}{\partial y}(a,b), \quad \frac{\partial g}{\partial x}(a,b), \quad \text{and} \quad \frac{\partial g}{\partial y}(a,b),$$

287

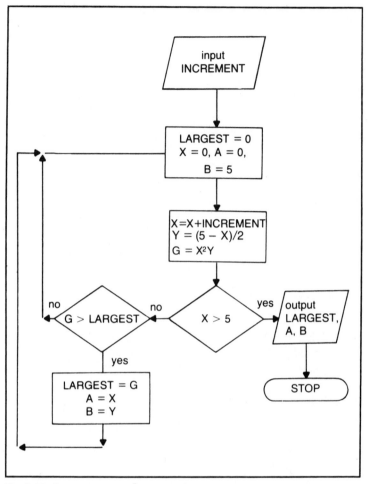

Fig. 5-43. Computing max x^2y subject to constraint.

as shown in the section on Directional Derivatives. It would then follow that an approximate value for λ would be the ratio of the two vector magnitudes,

$$\lambda = \frac{\sqrt{\left[\dfrac{\partial g}{\partial x}\,(a,b)\right]^2 + \left[\dfrac{\partial g}{\partial y}\,(a,b)\right]^2}}{\sqrt{\left[\dfrac{\partial f}{\partial x}\,(a,b)\right]^2 + \left[\dfrac{\partial f}{\partial y}\,(a,b)\right]^2}}$$

Unfortunately, all this estimation could lead to considerable error in the value for λ. If you desire a fairly accurate estimate for λ, very small increment values for x and for computing the partial derivatives would have to be used.

Finally, consider the situation in which the functions f and g have three variables, say x,y,z. Suppose you wish to maximize g(x,y,z) subject to f(x,y,z) = K for some constant K. As mentioned earlier, the technique of LaGrange multipliers would assert

$$\overrightarrow{\text{grad}}\ g(x,y,z) = \lambda\ \overrightarrow{\text{grad}}\ f(x,y,z)$$

where

$$\overrightarrow{\text{grad}}\ g(x,y,z) = \frac{\partial g}{\partial x}\ (x,y,z)i + \frac{\partial g}{\partial y}\ (x,y,z)j + \frac{\partial g}{\partial z}\ (x,y,z)k.$$

Assuming that you are either unfamiliar with partial derivatives or unable to solve the associated system of equations that involve λ, let's try to approximate the solution point (a,b,c). This means that f(a,b,c) = K and g(x,y,z) ≤ g(a,b,c) for all points (x,y,z) that belong to f. The programming method will involve letting the variables x and y run through nested loops. You will need to determine beforehand the intervals to which x and y belong.) Then z is determined from the side condition. This is the context for the next programming problem.

Exercises

1. A company's profit P is found using the equation P = xyz, where x,y,z are three variables that assume only positive values, but satisfy the constraint x + 3y + 4z = 108. Write a program that will approximate the maximum profit P and will approximate the values x,y,z that yield this maximum profit.
2. What is the value for the LaGrange multiplier λ for the preceding program problem?

EULER'S CONSTANT

Can you remember back to the good old days when slide rules rather than pocket calculators prevailed as the fastest means for computing products, quotients, and trigonometric values? Practically every high school senior enrolled in 4-th year math or physics carried one of these tools to class. Quite often it would be found dangling from the belt. Times have sure changed! Slide rules have now pretty much gone the way of hula hoops, sock hops, and model T's—namely OUT.

Nearly every slide rule had the two most important mathematical constants etched in a position corresponding to their approximate decimal value as shown in Fig. 5-44. These two constants are, of course, $\pi = 3.14159\cdots$ and $3 = 2.71828\cdots$. A few of the slide rules contained a third mathematical symbol, one that was probably never used by students. Furthermore, because of its deep mathematical meaning, the symobl carried no meaning for the students. This symbol, commonly denoted by γ and known as Euler's constant, represents a number with an approximate value of .577. Oddly enough, it remains today an open question as to whether γ is rational or irrational.

The formal definition of γ stems from the limiting value of a particular sequence of numbers. Thus you have

$$\gamma = \lim_{n \to \infty} \left[1 + \frac{1}{2} + \frac{1}{3} + \cdots + \frac{1}{n} - \ln(n) \right].$$

The sequence $1 + 1/2 + 1/3 + \cdots + 1/n$ is the n-th partial sum of the familiar harmonic series, which is known to diverge to infinity,

$$\sum_{k=1}^{\infty} 1/k = \infty.$$

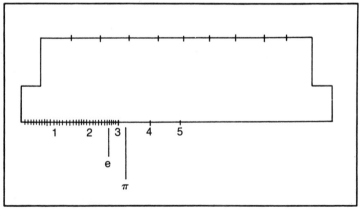

Fig. 5-44. A typical slide rule.

n	a_n
1	1.00000
2	.80685
3	.73472
4	.69704
5	.67390
6	.65824
7	.64695
8	.63842
9	.63174
10	.62638

Table 5-15. The Sequence Approximates to Euler's Constant.

But by subtracting ln(n) from the n-th partial sum, you will find, to your surprise that the sequence now converges. If you set a_n equal to this difference,

$$a_n = \sum_{k=1}^{n} 1/k - \ln(n)$$

the first few values of this sequence are listed in Table 5-15. It seems like the sequence $\{a_n\}$ might be a continually decreasing sequence. You could certainly obtain more values of a_n by having the computer perform the operations. The following is a sample program.

Program

```
100   N = 1
110   A(1) = 1
120   SUM = 1
130   FOR N = 2 TO 1000
140       N = N + 1
150       SUM = SUM + 1/N
160       A(N) = SUM - LN(N)
170       PRINT N,A(N)
180   NEXT N
```

Of course it is possible that these 1000 terms all decrease, while the terms after them increase (investigate $a_n = (1+1/n)^{n+.499}$).

One sound argument used for proving that $\{a_n\}$ always decreases is as follows: if you set $b_n = a_n - 1/n$, then

$$a_{n+1} - a_n = \frac{1}{n+1} - \ln(1 + 1/n)$$

and

$$b_{n+1} - b_n = \frac{1}{n} - \ln(1 + 1/n).$$

The elementary inequality, which you may want to prove,

$$\frac{1}{n+1} < \ln(1 + 1/n) < \frac{1}{n}$$

shows that $a_{n+1} < a_n$, and $b_{n+1} > b_n$ for all n. Thus, the sequence $\{a_n\}$ decreases and $\{b_n\}$ increases. Furthermore, $a_n > b_n$ implies that $\{a_n\}$ is bounded below by $b_1 = 0$. Hence $\{a_n\}$ must converge, and this limit is denoted by γ.

The importance of Euler's constant centers particularly around its relationship with some of the more powerful functions in applied mathematics. One of these is the gamma function, $\Gamma(x)$. The standard definition for $\Gamma(x)$ is the integral

$$\Gamma(x) = \int_0^\infty e^{-t} t^{x-1} dt$$

but there are several equivalent formulations. One is the infinite product [3]

$$\Gamma(x) = \frac{e^{-\gamma x}}{x} \prod_{i=1}^\infty \frac{e^{x/i}}{1 + x/i}$$

which is noteworthy because of the presence of Euler's constant. In studying the convexity of Γ, it is important to compute the logarithm of Γ,

$$\log \Gamma(x) = -\gamma x - \log(x) + \sum_{i=1}^\infty \left(\frac{x}{i} - \log\left(1 + \frac{x}{i}\right) \right).$$

This function is not only continuous for positive x, but it is also differentiable, with the derivative being given by

$$\frac{d}{dx}(\log(x)) = \frac{\Gamma'(x)}{\Gamma(x)}$$

$$= -\gamma - \frac{1}{x} \times \sum_{i=1}^{\infty}\left(\frac{1}{i} - \frac{1}{x+i}\right).$$

From this, substituting x = 1 gives

$$\frac{\Gamma'(1)}{\Gamma(1)} = -\gamma.$$

Since $\Gamma(1) = 1$, you have $\Gamma'(1) = -\gamma$! See Fig. 5-45 for the geometric interpretation of this.

Euler's constant can be shown to be related to the Bernoulli numbers B_i in the following manner [40]:

$$\gamma = \lim_{n\to\infty}\left[\sum_{k=1}^{n}\frac{1}{k} - \ln(n) - \frac{1}{2-n} + \frac{B_1}{2n^2} - \frac{B_2}{2n^4} + \cdots\right]$$

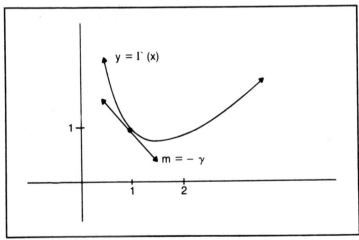

Fig. 5-45. Derivative of $\Gamma(x)$ at x = 1.

293

Since $B_1 = -1/2$ and $B_2 = 1/6$, you can approximate γ by the sequence b_n where

$$b_n = \sum_{k=1}^{n} \frac{1}{k} - \ln(n) - \frac{1}{2n} - \frac{1}{4n^2} - \frac{1}{12n^4} \, .$$

The first five terms in this sequence are $b_1 = .16667$, $b_2 = .48914$, $b_3 = .53924$, $b_4 = .55609$, and $b_5 = .56376$. You can see that these terms apparently converge to γ faster (as they well should) than $\{a_n\}$ does. In fact, the difference $|b_5 - \gamma|$ is less than .0135; while $|a_5 - \gamma|$ is approximately .0967. Furthermore, for $n = 10$, you have $|b_{10} - \gamma| \leq .0067$ while $|a_{10} - \gamma| \leq .0492$.

Finally, in an attempt to further understand how $\{a_n\}$ converges to γ, it might prove useful to express a_n as an integral. Because of the identities [66]

$$\frac{1}{n} = \int_0^{\infty} e^{-nx} dx$$

$$\ln(n) = \int_0^{\infty} \left(\frac{e^{-x} - e^{-nx}}{x} \right) dx$$

it follows that

$$a_n = \sum_{k=1}^{n} \frac{1}{k} - \ln(n)$$

$$= \sum_{k=1}^{n} \int_0^{\infty} e^{-kx} dx - \int_0^{\infty} \left(\frac{e^{-x} - e^{-nx}}{x} \right) dx$$

$$= \int_0^{\infty} \left(\frac{e^{-x} - e^{-(n+1)x}}{1 - e^{-x}} - \frac{e^{-x} - e^{-nx}}{x} \right) dx.$$

Thus,

$$\gamma = \lim_{n \to \infty} a_n = \int_0^{\infty} \left(\frac{e^{-x}}{1 - e^{-x}} - \frac{e^{-x}}{x} \right) dx$$

which means that if you wish to investigate the difference $a_n - \gamma$, it might prove beneficial to study the integral

$$A_n - \gamma = \int_0^\infty e^{-nx} \left(\frac{1}{x} - \frac{1}{e^x - 1} \right) dx.$$

An important result [36] concerning the difference $a_n - \gamma$ states that

$$\sum_{k=1}^n \frac{1}{k} - \ln(n) = \gamma + O(1/n).$$

It might be appropriate to explain this notation. The symbols O, and o, are often used to compare the values of two functions as the independent variable tends to some limiting value. Thus, to write f = O(g) as x →c means that there is some constant A for which f(x) = Ag(x) as x approaches c. For example, you could write the following:

1. $\sin(x) = O(x)$ as $x \to 0$ $\quad$ (A = 1)
2. $\sqrt{x^2 + 1} = O(x)$ as $x \to \infty$ $\quad$ (A = 1.1).

The smaller size o, when used as f = o(g), means that the quotient f/g has a limiting value of zero, as x → c. Two examples here would include:

3. $\sqrt{x} = o(x)$ $\quad$ as $x \to \infty$
4. $e^n = o(n^n)$ $\quad$ as $n \to \infty$.

The variable need not be a continuous variable, as in examples 1, 2, and 3. It could just as easily be a discrete variable as in 4. If you set functions f and g equal to the following,

$$f(n) = \sum_{k=1}^n \frac{1}{k} - \ln(n) - \gamma$$
$$g(n) = 1/n$$

the above result states that f = O(g) as n → ∞. To help verify this you should run a computer program to gather some figures that would help you determine a possible value for A. This is the essence of the following programming problem.

Exercises

1. If you set f(n) $= 1 + 1/2 + \cdots + 1/n - \ln(n) - \gamma$ (with $\gamma =$ .57721) and g(n) $= 1/n$, verify that f $= O(g)$ by obtaining a good approximate for A. This means you need to find A for which f(n) $\leq$ A(1/n) or nf(n) $\leq$ A. To this end, write a program that computes the values of a_n and f(n) for n $= 1,2,\cdots,100$. Print out these values for n, a_n, f(n) and nf(n) in tabular form, and see if you can't determine a good choice for A (nf(n) $\leq$ A).

2. Establish the inequality $\ln(1 + 1/n) < 1/n$ for n $= 1,2,\cdots$ by showing that f(x) > 0 for x ≥ 1 where f(x) $= (1/x) - \ln(1 + 1/x)$.

EULER'S METHOD

Much has been written on methods that can be used to solve particular classes of differential equations. In fact, complete texts are devoted to the subject. These methods are so many and varied that it is difficult, if not impossible, for anyone who is unaccustomed to handling differential equations to know where and how to get started solving an equation when one presents itself. It could prove more practical to learn some techniques for approximating the solution to a differential equation, especially in view of the fact that an explicit formula containing elementary functions and satisfying the equation need not exist. This is quite often what people do, particularly in the study of non-linear equations.

Assuming that a solution to a differential equation exists, it represents a locus of points in the plane. This locus of points forms a smooth, continuous curve. Throughout the remainder of this section, attention shall be focused on first-order differential equations of the form

$$\frac{dy}{dx} = f(x,y).$$

You don't need to concern yourself with higher order equations, at least not at this time, primarily because they can be reduced to a system of first-order equations.

One of the simplest techniques for approximating the solution to the above equation is known as Euler's method. This method allows you to obtain some rapid estimates to the solution, without having to do lengthy and messy computations. There are a few drawbacks to the method, as is usually the case with an approximating algorithm, but the method lends itself quite well to a more generalized and improved version [81].

To begin with, let y denote the exact solution to the equation

$$\frac{dy}{dx} = f(x,y)$$

and assume that a side condition is given; for example, the solution curve passes through the point (x_0, y_0). This is equivalent to saying $y(x_0) = y_0$. The symbol h shall represent a fixed, positive increment in x. Then define a sequence of x values by $x_1 = x_0 + h$, $x_2 = x_1 + h$, •••, $x_{n+1} = x_n + h$. In particular, on the interval $[x_0, x_1]$, the integral of $f(x,y)$ is $y(x_1) - y(x_0)$:

$$\int_{x_0}^{x_1} f(x,y)dx = \int_{x_0}^{x_1} \frac{dy}{dx} dx$$

$$= y(x_1) - y(x_0).$$

This means that the value of the function y at x_1 is given by

$$y(x_1) = y(x_0) + \int_{x_0}^{x_1} f(x,y)dx$$

$$= y_0 + \int_{x_0}^{x_1} f(x,y)dx.$$

Now here is where the first approximation takes place. Not knowing the exact behavior of f on the interval $[x_0, x_1]$, assume that it varies slowly (you can usually tell by inspecting the behavior of a particular f), which certainly depends heavily on the size of h. The function f is then replaced by the constant function of f evaluated at the left end point x_0. Thus

$$y(x_1) \doteq y_0 + \int_{x_0}^{x_1} f(x_0, y_0)dx$$

This expression simplifies to

$$y(x_1) \doteq y_0 + f(x_0, y_0)[x_1 - x_0]$$

297

$$= y_0 + hf(x_0,y_0).$$

Thus, we have

$$y_1 = y(x_1) = y(x_0 + h) \doteq y_0 + hf(x_0,y_0).$$

Having obtained y_1, you can then proceed to obtain function values y_2, y_3, ••• at x_2,x_3, ••• in a similar manner. This yields the formulas,

$$y_2 \doteq y_1 + hf(x_1,y_1)$$
$$y_3 \doteq y_2 + hf(x_2,y_2)$$
$$\bullet$$
$$\bullet$$
$$\bullet$$
$$y_{n+1} \doteq y_n + hf(x_n,y_n).$$

As an example, you can try this iteration scheme on a differential equation for which you know the explicit solution. You can compare the estimated function values with the exact function values and then obtain a feel for the accuracy of Euler's method.

Consider the equation $dy/dx = f(x,y) = xy$ along with the constraint $y(0) = 1$; thus $x_0 = 0$ and $y_0 = 1$. Set the increment $h = 0.2$, and using the above set of equations, you will be able to estimate the function values at $x = .2, .4, .6, .8,$ and 1.0. These values are listed in Table 5-16. Consequently the solution curve y not only passes through the point (0,1) but also comes close to passing through the five points (.2,1), (.4,1.04), (.6,1.12), (.8,1.26), and (1,1.46). You should be able to obtain better approximations by choosing a smaller value for h; say, $h = 0.1$. These values, rounded off to three digits past the decimal point, are given in Table 5-17.

Table 5-16. Employing Euler's Method on dy/dx= xy.

x_i	y_i
.2	$y_0 + hf(x_0,y_0) = 1 + .2(0)(1) = 1$
.4	$y_1 + hf(x_1,y_1) = 1 + .2(.2)(1) = 1.04$
.6	$y_2 + hf(x_2,y_2) = 1.04 + .2(.4)(1.04) = 1.1232$
.8	$y_3 + hf(x_3,y_3) = 1.1232 + .2(.6)(1.1232) = 1.257984$
1.0	$y_4 + hf(x_4,y_4) = 1.257984 + .2(.8)(1.257984) = 1.4592781$

x_i	y_i
.1	1.000
.2	1.010
.3	1.030
.4	1.061
.5	1.104
.6	1.159
.7	1.228
.8	1.314
.9	1.419
1.0	1.547

Table 5-17. A Smaller Increment h for Euler's Method.

The exact solution to this differential equation is $y = e^{x^2/2}$ because

$$\frac{dy}{dx} = xe^{x^2/2} = xy$$

and $y(0) = 1$. You can therefore make a comparison between the exact values for y_i and the approximated values, using both $h = 0.2$ and $h = 0.1$. These values are tabulated in Table 5-18.

It is apparent that $h = 0.1$ gives more accurate results than $h = 0.2$, and this is to be expected. But how much better are the values? It might be worthwhile to compare the differences and to compute the percentage of relative error, which is defined to be

Table 5-18. Comparing Approximate and Exact Solutions for dy/dx=xy.

	h = 0.2	h = 0.1	exact
x_i	y_i	y_i	y_i
.1		1.000	1.005
.2	1.000	1.010	1.020
.3		1.030	1.046
.4	1.040	1.061	1.083
.5		1.104	1.133
.6	1.123	1.159	1.197
.7		1.228	1.278
.8	1.258	1.314	1.377
.9		1.419	1.499
1.0	1.459	1.547	1.649

299

$$\frac{|\text{error}|}{\text{true value}} \times 100.$$

These differences are given in Table 5-19. It appears that by halving the increment h, the percentage of relative error is practically halved. On the other hand, for a fixed value of h, the error increases as you proceed over a larger range farther from the initial point x_0.

Euler's method is sometimes referred to as the method of tangent lines. This nomenclature applies because of the geometrical interpretation of the recursive relation $y_{n+1} = y_n + hf(x_n, y_n)$. To see

Table 5-19. Error and Relative Error for Solutions to dy/dx = xy.

x_i	h = 0.2 error	% rel. error
.1		
.2	.020	1.96
.3		
.4	.043	3.97
.5		
.6	.074	6.18
.7		
.8	.119	8.64
.9		
1.0	.190	11.52

x_i	h = 0.1 error	% rel. error
.1	.005	.50
.2	.010	.98
.3	.016	1.53
.4	.023	2.12
.5	.029	2.56
.6	.038	3.17
.7	.050	3.91
.8	.063	4.58
.9	.080	5.34
1.0	.102	6.19

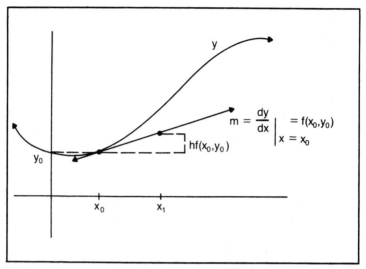

Fig. 5-46. Method of tangent lines.

this, first sketch in the solution curve y, which contains the point (x_0, y_0). The expression

$$hf(x_0, y_0) = h\left.\frac{dy}{dx}\right|_{x = x_0}$$

represents the vertical increase of the first approximate. This point (x_1, y_1) lies on the line tangent to y at (x_0, y_0) as shown in Fig. 5-46. Similarly, the expression

$$hf(x_1, y_1) = h \cdot \left.\frac{dy}{dx}\right|_{x = x_1}$$

represents the vertical increase of the second approximate. This point (x_2, y_2) lies on the line that passes through (x_1, y_1) and has the same direction as the line tangent to y at x_1 as shown in Fig. 5-47. The pattern continues with each point (x_n, y_n) lying on a line that is parallel to a corresponding tangent line to the solution curve y.

To summarize, the Euler method, though attractive because of its simplicity ($y_{n+1} = y_n + hf(x_n, y_n)$), is seldom used for serious calculations because of the quick buildup of error. Instead, an improvement of the method, known as the modified (or improved) Euler method, offers greater precision. This formula is given by,

$$y_{n+1} = y_n + \frac{1}{2}h\left[f(x_n,y_n) + f(x_{n+1},y^*_{n+1})\right]$$

where

$$y^*_{n+1} = y_n + hf(x_n,y_n).$$

If you consider the previous example of $dy/dx = f(x,y) = xy$, with $y(0) = 1$, then setting $h = 0.1$, you can compute the first few approximates y_i as follows:

$x_1 = 0.1$ $\quad y^*_1 = y_0 + hf(x_0,y_0) = 1 + .1(0)(1) = 1$

$$y_1 = y_0 + \frac{1}{2}h\left[f(x_0,y_0) + f(x_1,y^*_1)\right]$$
$$= 1 + (1/2)(.1)[0 + .1(1)]$$
$$= 1.005$$

$x_2 = 0.2$ $\quad y^*_2 = \frac{y_2}{2} y_1 + 1\ h f(x_1,y_1) = 1.005 + .1(.1)(1.005)$

$$= 1.01505$$

$$y_2 = y_1 + \quad h\left[f(x_1,y_1) + f(x_2,y_2^*)\right]$$
$$= 1.005 + (1/2)(.1)\left[(.1)(1.005) + (.2)(1.01505)\right]$$
$$= 1.0201755$$

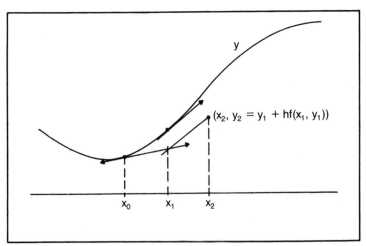

Fig. 5-47. The second approximate in Euler's method.

Table 5-20. Modified Euler Method, h = 0.1.

x_i	y_i	exact y_i	% rel. error
.1	1.0050	1.0050	0.0
.2	1.0202	1.0202	0.0
.3	1.0460	1.0460	0.0
.4	1.0832	1.0833	.01
.5	1.1331	1.1331	0.0
.6	1.1971	1.1972	.01
.7	1.2774	1.2776	.02
.8	1.3768	1.3771	.02
.9	1.4988	1.4993	.03
1.0	1.6479	1.6487	.05

$$x_3 = 0.3$$
$$y^*_3 = y_2 + hf(x_2,y_2) = 1.0201755 + .1(.2)(1.0201755)$$
$$= 1.040579$$
$$y_3 = y_2 + \frac{1}{2}h[f(x_2,y_2) + f(x_3,y^*_3)]$$
$$= 1.0201755 + (1/2)(.1)\ [(.2(1.0201755)$$
$$+ (.3)(1.040579)]$$
$$= 1.0459859$$

Notice that you must compute y^*_n before calculating y_n. These values and the next few approximations are listed in Table 5-20 along with the exact values, and percentage of relative errors. The degree of precision of the improved Euler method is quite obvious from this particular example.

Exercises

1. Write a program that will calculate the approximate values $y_1, y_2, \cdots, y_n$ for a differential equation of the form

$$\frac{dy}{dx} = f(x,y).$$

Assume, for simplicity, that f is always a linear function of the form $f(x,y) = Ax + By + C$, where the coefficients are real numbers. These coefficients will be fed into the program along with values for n and h. Print out the values y_1 that are computed using both Euler's method and the modified version. Test your program with $dy/dx = x^2/y$, $y(0) = 1$.

2. Suppose $dy/dx = y$ and that $y(0) = 1$. Set $h = 1/n$. What is a closed expression for y_n? Evaluate $\lim y_n$.

Chapter 6
Probability and Statistics

A recent survey conducted by the Curriculum Committee of the Ohio Section of the Mathematics Association of America (M.A.A.) provided strong evidence that today's college graduates, particularly those with major interests in the mathematical sciences or business, deem a course in probability and statistics one of the most beneficial of courses possible. This seems to be logical because of the increased importance placed on applications today, and also because there are so many businesses and fields of endeavor where empirical data is the only data available. This data is used to test critical hypotheses.

You are all familiar with the role that probability plays in helping determine the likelihood that a specific event will occur. It may be that you are interested in whether or not team A will win the ballgame, person B will achieve a score within a particular range, company C will have profits in excess of a specific amount, or a mathematical equation has D number of solutions, or whether an infinite series converges or diverges. The laws that govern probability can be either remarkably simple or excruciatingly complex, but in either case, they must be accurately applied in order for you to gather correct information. In some of the applications that appear in this final chapter I will try to simulate a situation with a mathematical model, and by repeated computer trials arrive at a numerical value that closely approximates the true probability. The section on Buffon's Needle Problem most vividly demonstrates this programming technique.

Statistics, like mathematics, is a universal language of the sciences. Careful use of statistical methods enables you to describe the findings of scientific research or business surveys and make decisions and estimations. Essentially, statistics is the science of collecting, classifying, presenting, and interpreting numerical data. Statistics asks you to draw a sample, describe the sample, and then make an inference about the population in general based on the information found in the sample. Complex mathematical equations are used to describe some of the important sampling distributions. These include the binomial distribution, the normal distribution, the Poisson distribution, and the F and Student's t distribution. These distributions help to describe the behavior of many of the discrete random variables encountered in most fields of applications.

EXPECTED VALUE

Often, problems of interest are introduced by the words, "What is the expected outcome of," or "What is the expected value of." These words are commonplace enough so that nearly everyone has a sufficiently sound idea of what the problem is really about. Suppose you are doing some carpentry work. To estimate an answer to, "What is the expected number of hammer poundings needed to completely drive the nail into the board" would probably present little or no difficulty. You might respond with an answer of seven or eight, and no one would argue. On the other hand, a question like, "In one roll of a single die, what number can you expect to get" could evoke several answers. Most typical among these are "You can expect any of the numbers 1,2,3,4,5,6 since they are all equally likely," or "You can expect 3.5 since that is the average of the six numbers." The point here is that it is not difficult to conjure up a situation where understanding the problem is somewhat of a task and determining the solution is even more so. This leads to the considerations of the statistical concept of the expected value: the expected value of a discrete random variable.

The expected value of a discrete random variable X, with probability function p(x), is denoted by E(X) and defined by

$$E(X) = \sum_{i=1}^{n} x_i p(x_i)$$

where the summation extends over all the various values that X can assume.

The expected value of a random variable may be likened to a measure of position for the probability distribution. It is simply a weighted mean of the possible outcomes of X, with the probability values p(x) used as weights. It is for this reason that E(X) is often called the mean of the probability distribution of X, or simply, the mean of X. The expected value of a random variable is often denoted by μ, which corresponds to the notation for the mean of a population. Thus, quite obviously, the value of E(X) may be a number that does not correspond to any of the possible outcomes of X. To illustrate, consider the event of rolling a single die. The random variable X, which represents the number appearing on the rolled die, can assume the values 1,2,3,4,5, or 6. Each value occurs with equal probability of 1/6, so $p(x_i = i) = 1/6$. Thus, the expected value is

$$
\begin{aligned}
E(X) &= \Sigma\, x_i p(x_i) \\
&= \Sigma\, i(1/6) \\
&= 1\,(1/6) + 2(1/6) + 3(1/6) + \cdots + 6(1/6) \\
&= 3.5.
\end{aligned}
$$

Furthermore, suppose you had a handful of 20 dice. If you rolled all 20 of these dice, and added up all the numbers showing, what sum would you expect? It would have to be 70 (20•3.5).

It is important to mention a few properties of the expectation operator, E( ). For if X and Y denote two random variables, and a and b denote constants, it is possible to establish the following:

a. $E(a) = a$
b. $E(aX) = aE(X)$
c. $E(a + X) = a + E(X)$
d. $E(aX + bY) = aE(X) + bE(Y)$

Property (b) was used above (with a = 20) to compute the expected sum on the 20 dice. In fact, property (b) can be quickly established by noting that

$$
\begin{aligned}
E(aX) &= \Sigma\, (ax_i)p(ax_i) \\
&= \Sigma\, (ax_i)p(x_i) \\
&= a\,\Sigma\, x_i p(x_i) \\
&= aE(X).
\end{aligned}
$$

You may wish to try to establish the remaining three properties using a similar argument.

It is important (because this applies to so many real world problems) to mention that if the random variable X is continuous, as opposed to discrete, the expected value is defined as the integral

$$E(X) = \int_{-\infty}^{\infty} xp(x)dx.$$

If the shelf life X of a particular loaf of bread in weeks has the probability function $p(x) = (10 - x)/50$ for $0 \leq x \leq 10$, then the expected shelf life of a loaf of this bread is 10/3 weeks,

$$E(X) = \int_{0}^{10} x(10-x)/50 \, dx$$

$$= 10/3.$$

Let us return to the discrete random variable so you can consider two particularly delightful problems dealing with expected value. The first is as [64] follows. Assume you are required to throw a pair of dice and obtain a total of 5 or a 5 on at least one of the dice. What is the expected number of throws required for this to occur? This is a problem that could be of interest to the serious backgammon player. To analyze this problem, note there are 15 different ways the dice could either give a total of 5 or show at least one single 5:

(1,4),	(2,3),	(3,2),	(4,1),	(5,1)
(5,2),	(5,3),	(5,4),	(5,5),	(5,6)
(1,5),	(2,5),	(3,5),	(4,5),	(6,5).

Assuming equal probabilities for the 36 possible throws of a pair of dice, the probability of rolling the desired event is 15/36, or 5/12. Set $p = 5/12$. Use the random variable X to represent the number of throws needed before a 5 or total of 5, is reached. Thus $p(x=1) = 5/12$ while $p(x=2) = (7/12)(5/12)$ and, continuing, you get the probabilities listed in Table 6-1. The expected value E(X) is the infinite sum

x	p(x)
1	5/12
2	(5/12)(7/12)
3	(5/12)(7/12)^2
4	(5/12(7/12)^3
5	(5/12(7/12)^4

Table 6-1. Some Individual Probabilities.

$$\frac{5}{12} + \frac{5}{12}\left[\frac{7}{12}\right] + \frac{5}{12}\left[\frac{7}{12}\right]^2 + \cdots$$

which sums to 2.4. Thus, on the average, the desired event will happen on the 2.4-th roll of the dice.

To run a computer simulation of these throws, you can proceed as follows. First input NTRIALS, which will represent the number of trials or games played. Each game will consist of throwing the two die, DIE1 and DIE2, until either a 5 or sum of 5 is obtained. The variable THROW will keep track of how many throws are necessary for each game, and the variable TOTAL will keep a running sum of the values of THROW. After the last game, when NGAMES = NTRIALS, the value of TOTAL/NTRIALS should be a close approximate to E(X) = 2.4. The flowchart shown in Fig. 6-1 outlines these steps in greater detail.

Suppose you alter this problem slightly and ask for the expected number of throws before either a sum of K or an individual die showing a K is reached. Can you make the appropriate change in the program?

The second problem states that if numbers are randomly chosen (with replacement) from the unit interval [0,1], the expected number of selections necessary until the sum of the chosen numbers first exceeds one is e = 2.718•••,[72]. This is a rather striking result: to have the number e appear in so simple a problem is quite surprising. But, on the other hand, choosing numbers at random is not an easy task, especially if the selections are truly random and not pseudorandom. Nor is it the kind of problem that lends itself to a rational (as opposed to an irrational or transcendental) solution.

To illustrate what this problem is all about, start selecting numbers at random, numbers that lie between 0 and 1. Select numbers only until their sum first exceeds one. Quite likely this would happen after, say, choosing three numbers. If you repeat this

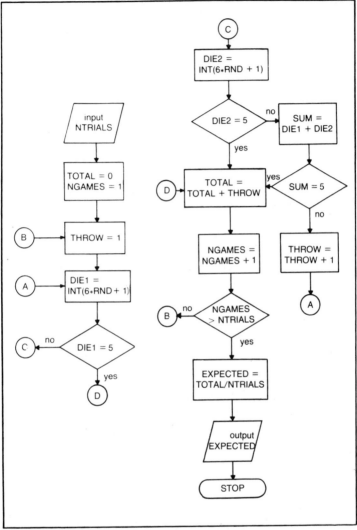

Fig. 6-1. Flowchart for computing expected value.

process, say seven more times, you might get the data listed in Table 6-2.

On the average, the number of selections needed before the sum of the random numbers first exceeds one is $22/8 = 2.75$. This is close to $e = 2.718\cdots$, and in fact, as the number of trials continues to increase, the average number of selections will converge to e. That is the meaning of saying the expected number of selections necessary is e.

To simulate this process on a computer, you might decide to run N trials, where, say, N is no bigger than 1000. The result of each trial (i.e., the number of selections necessary) can be stored in an array, TRIAL( ), and then after the N trials are over, you will just sum up TRIAL(1) + TRIAL(2) + $\bullet\bullet\bullet$ + TRIAL(N) and divide by N to get your approximate for the expected value. To assist you, the following program is presented.

Program

```
110   REM PROGRAM ON EXPECTED VALUE
115   DIM TRIAL (1000)
120   INPUT "THE VALUE OF N IS"; N
125   PRINT PRINT
130   PRINT TAB(5); "AN EXPECTED VALUE
      PROBLEM"
135   PRINT PRINT
140   FOR NTRIALS = 1 TO N
145       SUM = 0
150       FOR PICKS = 1 TO 50
155           SUM = SUM + RND
160            IF SUM   1 GO TO 180
165       NEXT PICKS
170       PRINT "NOT ENOUGH SELECTIONS"
175       GO TO 225
180       TRIAL(NTRIAL) = PICKS
185   NEXT NTRIALS
```

Trial	Number of Selections
1	3
2	3
3	2
4	2
5	3
6	2
7	4
8	3

Table 6-2. Choosing Random Numbers Until the Sum Exceeds One.

```
190   REM NOW TO COMPUTE THE AVERAGE
195   AVERAGE = 0
200   FOR NTRIALS = 1 TO N
205         AVERAGE = AVERAGE + TRIAL (NTRIALS)
210   NEXT NTRIALS
215   AVERAGE = AVERAGE/N
220   PRINT "THE AVERAGE IS"; AVERAGE
225   STOP
230   END
```

Exercises

1. Write a program that simulates the following two experiments, and collects enough data so that you can estimate the desired expected value.

   Experiment 1. You wish to roll N dice ($N = 1,2,\cdots,10$) with hopes of obtaining either a total of K ($K = 1,2,\cdots,100$), or a K on at least one of the dice. What is the expected number of rolls necessary to do this?

   Experiment 2. For each value of BOUND fed into the program, you are interested in the expected number of selections (with replacement) of random numbers from the unit interval before their sum first exceeds BOUND. It has been conjectured that this expected number, E(X), is given by

   $$E(X) = 2 * BOUND + (e - 2).$$

   What do you think?

2. What is the expected number of throws of a pair of dice in order to obtain a sum total of 4, or a 4 on at least one of the dice?

MEASURES OF CENTRAL TENDENCY

Every text on statistics contains an introductory chapter on the measures of central tendency, specifically the mean, the median, and the mode. It is obvious that these measures are important because they help to interpret and analyze data. Each of these three measures is designed to provide a different interpretation of the "average" of the data scores. For instance, if the average biology exam score was 78, would you know what that means? What about the average price of a new home in the United States during 1980?

Or the average life of a particular type of John Deere farm machinery?

It could very well be that all three uses of "average" from above are different: that the first average would make the most sense if the mean (the arithmetic mean) were used; that the second average is just a stand-in for the median measure; and the third would be in reference to the mode. It doesn't have to be this way: the three notions of average could carry different meanings, but it could very easily be as described above. Either way, the true meaning is ambiguous and uncertain the way it stands.

To proceed, recall how each of these fundamental concepts is defined. If your data set consists entirely of the scores $x_1, x_2, \cdots, x_n$, the mean, denoted by $\overline{x}$, is given by

$$\overline{x} = \frac{1}{n} \sum_{i=1}^{n} x_i.$$

Thus, the mean score for the values 2, 6, and 13 is 7 ($\overline{x} = 7$). Now, if the data scores are arranged in ascending order so that $x_1 \leq x_2 \leq x_3 \leq \cdots \leq x_n$, then the median score (also known as the 50th percentile, or the second quartile), denoted by x_{md}, is that score which is located in the middle. Two cases are customarily considered here. If there are an odd number ($n = 2k + 1$) of scores, then the median x_{md} is the $(k+1)$st score, so $x_{md} = x_{k+1}$. This means that the median is located in the position corresponding to the $(k+1)$st score. On the other hand, if there are an even number ($n = 2k$) of scores, the median is that numerical value located halfway between x_k and x_{k+1}; so $x_{md} = (1/2)(x_k + x_{k+1})$. This means that the median is equal to the mean of the two-point set $\{x_k, x_{k+1}\}$. To illustrate, the median for the data, 1, 3, 4, 9, 11 is 4, while the median for the set 1, 3, 4, 9, is 3.5.

The mode, denoted x_{mo}, is simply that score which appears most often. So if the data set consists of 1, 1, 2, 3, 4, 7, the mode is 1 ($x_{mo} = 1$). It could very well be that there are several modes (thus the mode score need not be unique, unlike the mean and median) such as when the data includes 1, 1, 2, 2, 3, 4, 5. This particular distribution is bimodal because there are two modes, 1 and 2. Clearly you can construct data with as many modes as you want—situation that is not very interesting or informative. Even though the term mode is not that common in everyday life (although neither is mean or median), its application is duly hidden behind such familiar phrases as usual, customary, or prevalent. Your usual dental bill for a cleaning might be $21: this typifies the interpretation of mode.

Now look at a few characteristic properties of these three measures of central tendency.

Property 1. The sum of the deviations of the x_i values from their mean $\overline{x}$ is zero. Thus,

$$\sum_{i=1}^{n} (x_i - \overline{x}) = 0$$

and this can be represented graphically, as shown in Fig. 6-2, by plotting the points (i, x_i) in the plane for each i and noting that the sum of the distances of the points below the line $y = \overline{x}$ is the same as the sum of the distances above the line.

Property 2. The mean $\overline{x}$ tends to minimize a certain mathematical expression concerned with the dispersion of values in the data set. More specifically, if A is any given number, then the finite sum S

$$S = \sum_{i=1}^{n} (x_i - A)^2$$

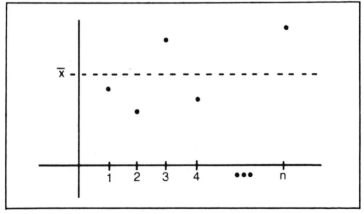

Fig. 6-2. A graphical interpretation of the mean.

assumes a minimum value only when $A = \overline{x}$. Thus when the data set contains, say, five elements ($n = 5$), with $x_1 = 1$, $x_2 = 3$, $x_3 = 4$, $x_4 = 9$, $x_5 = 13$, Table 6-3 lists some of the values for S according to different values selected for A. Note that the least value for S occurs when $A = 6$. This is when $A = \overline{x}$!

Table 6-3.
Corresponding Values of A and S.

A	S
2	176
4	116
6	96
8	116
10	176
12	276

Property 3. For large data sets where the values do not tend to repeat themselves extensively, the median can be interpreted to indicate that about 50% of the scores are smaller than the median, and about 50% are larger. This is one reason why the median is a measure of central tendency.

Property 4. The median x_{md} also minimizes a certain mathematical expression concerned with the variability of scores. Namely, if A is any given number, then the sum S,

$$S = \sum_{i=1}^{n} |x_i - A|$$

attains its minimum when $A = x_{md}$. The median is said to minimize the sum of the absolute deviations of the data scores. Do you know how to interpret this graphically?

Property 5. For any of the six possible orderings of the mean, median, and mode (such as $\overline{x} < x_{md} < x_{mo}$, or $x_{mo} < \overline{x} < x_{md}$) there is some data set for which this ordering holds. For example, the arrangement $x_{mo} < \overline{x} < x_{md}$

would be fulfilled with the data 1, 1, 1, 2, 2, 3, 3. The arrangement $\overline{x} < x_{md} < x_{mo}$ would hold for the set 0, 2, 3, 3. It might make for an interesting program to construct a data set with whichever of the six possible orderings desired. You might even try imposing the further condition that the size of the data set be given initially.

Property 6. Both the mean and the median appear in various formulas measuring either the variation of the scores (standard deviation, coefficient of variation) or the skewness (Pearson's coefficient of skewness). An extremely important bound on the difference between the mean and the median asserts [10,25] that their difference is at most one standard deviation. Thus, if the data includes 1, 1, 2, 3, 3, 4, 5, 7, 8, 10 then $\overline{x} = 4.4$, $x_{md} = 3.5$. Their absolute difference is $| \overline{x} - x_{md} | = .9$, which is less than the standard deviation of 2.91. Sometimes the bound may not necessarily be a good one, but it is somewhat surprising to find that these three important statistical measures can be so elegantly related.

You will want to use a computer for the computation of the values of the mean, median and mode. To this end, suppose n values x_1, x_2, $\bullet\bullet\bullet$, x_n, are given. How do you proceed if you wish to compute $\overline{x}$, x_{md}, and x_{mo}? Are the three computations essentially independent, meaning that you need three separate and unrelated subprograms? Or would it be better to compute one of the measures first, and then use that to determine another? There is obviously very little problem in computing the mean. By putting the values in an array X of one dimension of at most, say, 1000 elements (more than you will probably need), you could proceed as follows.

Program

```
110   DIM X(1000)
120   INPUT N
130   MAT INPUT X
140   SUM = 0
150   FOR I = 1 TO N
160       SUM = SUM + X(I)
170   NEXT I
```

```
180   MEAN = SUM/N
190   PRINT "THE MEAN IS"; MEAN
200   STOP
```

The real problem comes in trying to compute the median. It is necessary here to arrange the terms in order, say in increasing order. This means there will be a rearrangement so that

$$y_1 \leq y_2 \leq y_3 \leq \cdots \leq y_n$$

and each y_j term is equal to a different one of the x_i terms. In particular, y_1 is the smallest of the x_i terms, and y_n is the largest. There are plenty of different methods for carrying out the remainder of this rearrangement. Two of the methods will presently be outlined, and perhaps you can devise more.

Method 1. Begin by searching through x_1, x_2, $\cdots$, x_n for the smallest term, which will be called x_i. Thus $x_1 \leq x_1$, $x_i \leq x_2$, $\cdots$, $x_i \leq x_n$. Note that it may be possible for several terms to have the same minimum value. In this case the one with the smallest subscript will be chosen. Set $y_1 = x_i$. Then consider the new sequence of $n-1$ terms, $\{x_1, x_2, \cdots, x_n\}$ $- \{x\}$, or $\{x_1, x_2, \cdots, x_{i-1}, x_{i+1}, x_{i+2}, \cdots, x_n\}$. From this sequence select the smallest term, call it x. Again there may be several terms with this value. Set $y_2 = x$. Then consider the new sequence of n-2 terms $\{x_1, x_2, \cdots, x_n\} - \{x_i, x_j\}$, select the smallest term and give its value to y_3. The process continues for n-3 more steps. This searching method presents some difficulty in the search through the new sequences, mainly because it is best to relabel the new sequence each time, and this chore is somewhat tedious.

Method 2. This is the so-called bubble method. Starting with the sequence $x_1, x_2, \cdots, x_n$, compare the first two terms, x_1 and x_2, placing the smaller term first and relabeling it. If by chance x_2 were smaller than x_1, your program would read as follows.

```
110   IF X(1) ≤ X(2) GO TO 150
120   A = X(1)
130   X(1) = X(2)
```

```
140   X(2) = A
150   - - - - -
```

Then you would compare the second and the third terms, x_2 and x_3, placing the smaller term before the other. The process continues until the last two terms are compared and the smaller one placed first. This sequence of n-1 steps completes phase 1 of the entire procedure, and what it really accomplishes is to set the largest element in the data set equal to x_n (in the n-th position. To illustrate phase 1, if the data is 2, 7, 3, 1, and 4, the n-1 = 5-1 = 4 steps would be:

first step: $x_1 = 2$, $x_2 = 7$, $x_3 = 3$, $x_4 = 1$, $x_5 = 4$
second step: $x_1 = 2$, $x_2 = 3$, $x_3 = 7$, $x_4 = 1$, $x_5 = 4$
third step: $x_1 = 2$, $x_2 = 3$, $x_3 = 1$, $x_4 = 7$, $x_5 = 4$
fourth step: $x_1 = 2$, $x_2 = 3$, $x_3 = 1$, $x_4 = 4$, $x_5 = 7$.

Notice that x_5 is now equal to the largest element. Now that you have a new arrangement of $x_1, x_2, \bullet\bullet\bullet, x_n$, you can begin phase 2, by first comparing the first two terms, and relabeling if necessary so the smaller term is first. Then, as in phase 1, you will do the same with the second and third terms, then the third and fourth terms, and so on up to terms x_{n-2} and x_{n-1}. Phase 2 consists of a sequence of n−2 steps, and for the example that was started above, these 5-2=3 steps would be:

first step: $x_1 = 2$, $x_2 = 3$, $x_3 = 1$, $x_4 = 4$, $x_5 = 7$
second step: $x_1 = 2$, $x_2 = 1$, $x_3 = 3$, $x_4 = 4$, $x_5 = 7$
third step: $x_1 = 2$, $x_2 = 1$, $x_3 = 3$, $x_4 = 4$, $x_5 = 7$.

The completion of phase 2 guarantees us that the next to largest element is located in the next to last position, x_{n-1}, in the array. Phase 3, similar to the others, will be an n−3 step rearranging of the first n−2 terms. Upon its completion the third largest term will be assigned to x_{n-2}. In the example, these 2 steps would be:

first step: $x_1 = 1$, $x_2 = 2$, $x_3 = 3$, $x_4 = 4$, $x_5 = 7$
second step: $x_1 = 1$, $x_2 = 2$, $x_3 = 3$, $x_4 = 4$, $x_5 = 7$

and at this stage the data has been completely ordered. The program might (depending on what conditional statements you write) continue on with phase 4, phase 5, $\bullet\bullet\bullet$, phase $(n-1)$, because after phase $(n-1)$, you can be certain that the data is rearranged in ascending order. As a sidelight, it could be very significant, especially if n is large, to have an estimate as to the expected number of phases needed before the data is completely rearranged.

Determining the median is a simple task once the scores are ordered. If n is an odd number, then $x_{md} = x_{(n+1)/2}$. But if n is even (n $= 2k$), then $x_{md} = (1/2)(x_k + x_{k+1})$.

It is generally believed to be easier to compute the mode once the data has been ordered. Since the data is now in the form

$$x_1 \leq x_2 \leq x_3 \leq \bullet\bullet\bullet \leq x_n$$

you really wish to separate it into disjointed classes where the elements in each set are all equal to each other. In the first class will be all the terms (say i_1 of them) equal to $x_1 : x_1 = x_2 = \bullet\bullet\bullet = x_{i_1}$. The second class will be the next batch of i_2 terms, all equal to $x_{i_1} + 1$: $x_{i_1} + 1 \ x_{i_1} + 2 = \bullet\bullet\bullet = x_{i_1} + 1$, and so on. This time, if your data set contains the following 10 scores, rearranged in ascending order as 1,1,2,2,3,3,3,4,5,5 you would divide this into five classes.

$$\text{class(1): 1, 1}$$
$$\text{class(2): 2, 2}$$
$$\text{class(3): 3, 3, 3}$$
$$\text{class(4): 4}$$
$$\text{class(5): 5, 5}$$

and count the number of elements in each class,

$$\text{count(1) = 2}$$
$$\text{count(2) = 2}$$
$$\text{count(3) = 3}$$
$$\text{count(4) = 1}$$
$$\text{count(5) = 2.}$$

Since count(3) is the largest of these five numbers, the mode for the 10 point data set is x_7, or alternately

$$x_{mo} = x_{[count(1) + count(2) + count(3)]}.$$

I hope you can observe the generalization.

Exercises

1. Write a program that computes the mean, median, and mode for N given data scores. Test your program with the following two sets of values.

    1. 4, 2, 3, 3, 5, 8, 6, 6, 3, 1
    2. 5, 8, 8, 3, 4, 4, 1, 7, 2

2. Construct a five point data set in which the three measures of central tendency are related by $x_{md} < \bar{x} < x_{mo}$.

MEANS

You are now very familiar with some of the more important measures of central tendency, namely the mean, median and mode. Of these three, the mean is not only the most widely known and used, but it is also the term that has been the most widely generalized to cover alternative concepts. The following discussion centers in on some of the more popular adaptions of the mean.

First of all, the mean as used in introductory statistics is simply an abbreviation for the arithmetic mean, denoted by AM.

$$AM = \frac{1}{n} \sum_{i=1}^{n} x_i$$

The adjective arithmetic is used because of the addition involved in the formula, addition being the most basic and fundamental of all the binary operations.

Next is the notion of the geometric mean, denoted by GM. Here you consider the product, as opposed to sum, of all the terms in the data set. In an effort to arrive at some figure having something to do with an average, you then take the n-th root of the product. Thus you have

$$GM = \sqrt[n]{x_1 x_2 \cdots x_n}.$$

To illustrate, if your data set contains the five numbers 3,4,5,6,7, the two means are:

$$AM = 5$$
$$GM = \sqrt[5]{3 \cdot 4 \cdot 5 \cdot 6 \cdot 7} = \sqrt[5]{2520} = 4.789.$$

These two means will always have different values except in the special case when all the scores in the data set are the same. For if $x_1 = x_2 = \cdots = x_n = a$, then $AM = (1/n)(a + a + \cdots + a) = (1/n)(na) = a$ and $GM = (a^n)^{1/n} = a$, so $GM = AM$. Quite often you will find an exercise in a book that requires you to prove that the arithmetic mean AM of two positive scores is at least as great as the geometric mean, $AM \geq GM$. Equivalently this would read

$$\frac{x_1 + x_2}{2} \geq (x_1 x_2)^{1/2}$$

which is also equivalent to $(x_1 + x_2) \geq 2\sqrt{x_1 x_2}$. When you square both sides, you get $x_1^2 + 2x_1 x_2 + x_2^2 \geq 4x_1 x_2$, or $x_1^2 - 2x_1 x_2 + x_2^2 \geq 0$. Since this last inequality is always true ($(x_1 - x_2)^2 \geq 0$), this implies that $AM \geq GM$ for any two positive terms. What is perhaps more remarkable is that $AM \geq GM$ irregardless of the number of positive terms [80]. A rather ingenious proof of this remarkable inequality appears in [16]. Interesting enough, the same inequality holds true for the generalized case of weighted arithmetic and geometric means [9,62]. Furthermore, note that if y_n is the geometric mean of the elements $x_1, x_2, \cdots, x_n$, the geometric mean of their reciprocals is $1/y_n$.

The geometric mean is functional in one sense in that it can furnish an "average" salary over a period of n years; an average that is obtained from the initial salary and the yearly salary increases. For example, suppose your salary one year was $10,000 and then for the next five years your salaries were $11,000, $12,400, $14,000, $16,000 and $17,500. This means the percentage increases were, respectively, 10%, 12.727%, 12.903%, 14.285% and 9.375%. Over the course of the six years, you could say that the average salary was $13483.33 – the arithmetic mean – and this would imply that the total wages earned was $80,900 = 6($13483.33): an average yearly raise of 11.858%. Now let us see how the geometric mean could be applied to this setting. The first salary increase of 10% means that the new salary was 1.10 times as great as the original salary. The following salary of \$12,400 was 1.12727 times as great as the \$11,000 salary. The remaining three salaries were 1.12903, 1.14285 and 1.09375 times as great as each previous salary. The geometric mean of these five ratios 1.10, 1.12727, 1.12903, 1.14285, and 1.09375 is

$$GM = \sqrt[5]{1.7499813} = 1.118424.$$

How does this number 1.118424 fit into the scheme of things from above? What it means is that if your salary each year increased to 1.118424 times as much as the previous year, at the end of the sixth year your salary would be approximately \$17,500—just as it was with the five variable increases. The results in Table 6-4 bear this out. Thus, if two people start out with the same initial salary, and if one receives raises of 10%, 12.727%, 12.903%, 14.285% and 9.375% and the other receives raises that are all 11.8424%, the two would be receiving the same salary during this last year, although probably not the same total pay during the six year interval.

Table 6-4. Salaries with Yearly Increases of 1.118424.

Year	Salary		
1	10000		
2	1.118424 × 10000	=	11184.24
3	1.118424 × 11184.24	=	12508.72
4	1.118424 × 12508.72	=	13990.05
5	1.118424 × 13990.05	=	15646.81
6	1.118424 × 15646.81	=	17499.77

Another prevalent generalization of the mean is the harmonic mean, denoted by HM. The harmonic mean of a set of scores $x_1, x_2, \cdots, x_n$ is the reciprocal of the arithmetic mean of the reciprocals of the individual scores; thus

$$HM = \cfrac{n}{\cfrac{1}{x_1} + \cfrac{1}{x_2} + \cdots + \cfrac{1}{x_n}}$$

The data scores are not allowed to include either zero or negative values, this prevents division by zero. To illustrate, the harmonic mean of the three scores 2, 3, and 4 is 2.769, where

$$\frac{3}{\frac{1}{2} + \frac{1}{3} + \frac{1}{4}} = \frac{36}{13} = 2.769.$$

As with its two predecessors, the harmonic mean will equal the geometric mean (or arithmetic mean) only if all the data scores are the same:

$$HM = \frac{n}{\frac{1}{a} + \frac{1}{a} + \cdots + \frac{1}{a}} = \frac{n}{\frac{n}{a}} = a.$$

Otherwise the value of HM is always less than GM [62].

This particular blend of generality proves to be especially useful as an averaging tool when the data scores represent rates of change. For instance, on a recent drive to Cleveland, a driver traveled at 60 mph for the first 30 miles and then 40 mph for the next 30 miles. You need to determine an average speed for the entire trip. How about 50 mph? You would be tempted to compute the arithmetic mean, AM = (1/2)(60 + 40), and conclude that 50 mph represented the average speed. This is inaccurate because the total distance traveled was 30 + 30 = 60 miles, the total time consumed was .5 + .75 = 1.25 hours; thus the average speed was 60/1.25 = 48 mph. The harmonic mean for the two scores 60 and 40 is precisely this figure,

$$HM = \frac{2}{\frac{1}{60} + \frac{1}{40}} = 48.$$

It is important to note here that the harmonic mean formula is effective and functional because the two rates were maintained over equal distances (30 miles each). What would happen if the distances were unequal?

A fourth type of mean is the logarithmic mean, LM. Chemical engineers often find use for this measure, which is radically different from the three others in that only two data scores (both positive) are allowed. If these two scores are x_1 and x_2, the value of the log-mean is defined to be

$$LM = \frac{x_1 - x_2}{\ln(x_1/x_2)}$$

where ln denotes the natural logarithm. The log-mean can be viewed as a function of two variables, the two data scores. It is common to denote this by writing $LM(x_1,x_2)$. Thus, the logarithmic mean of 1 and 2 is

$$LM(1,2) = \frac{1}{\ln(2)}$$

and the logarithmic mean of e^7 and e^{11} is

$$LM(e^7,e^{11}) = \frac{e^{11} - e^7}{\ln(e^4)} = \frac{e^{11} - e^7}{4}.$$

From a mathematical standpoint, it is obvious that if $x_1 = x_2$, the log-mean is undefined because of the division by zero, but what would be the value of $LM(x_1,x_2)$ if x_1 and x_2 were almost equal? Since the value of $LM(x_1,x_2)$ is sandwiched between x_1 and x_2 (see if you can prove this), it follows that if x_1 is approximately equal to x_2, $LM(x_1,x_2)$ is approximately equal to x_2. In fact, because of the above sandwiching, the following limit is immediately apparent:

$$\lim_{x_1 \to x_2} LM(x_1,x_2) = x_2.$$

Furthermore, the function is symmetric in the sense that $LM(x_1,x_2) = LM(x_2,x_1)$.

The most interesting relationship involving LM is in its relationship to the arithmetic and geometric mean. Here, with a data set of two, you have

$$GM < LM < AM.$$

So if x_1 and x_2 are different positive numbers, then

$$\sqrt{x_1 x_2} < \frac{x_1 - x_2}{\ln(x_1/x_2)} < \frac{x_1 + x_2}{2} .$$

Don't forget that the harmonic mean is less than the geometric mean! Just to get an idea as to the value of these mean measures, consider the data presented in Table 6-5.

Table 6-5. Sample Data on the Four Mean Measures.

x_1	x_2	HM	GM	LM	AM
2	3	2.4	2.449	2.466	2.5
2	4	2.666	2.828	2.885	3.0
2	5	2.857	3.162	3.274	3.5
2	6	3.0	3.464	3.641	4.0
2	7	3.111	3.741	3.991	4.5
2	8	3.2	4.0	4.328	5.0

Examining this table might make you ask questions concerning either the distribution of values for a particular measure or the variation between the different measures. Do the values of AM increase at the fastest rate? Faster than LM? If you compute the arithmetic mean of the two values AM and HM, is the value closest to GM or LM? What if you calculate the log-mean of AM and HM? This last question can be reformulated using symbols as follows: if $x_1 \neq x_2$ are positive, is the value of

$$\frac{\dfrac{x_1 + x_2}{2} - \dfrac{2}{\dfrac{1}{x_1} + \dfrac{1}{x_2}}}{\ln\left[(x_1+x_2)/2\right] / \left[2/(1/x_1 + 1/x_2)\right]}$$

best approximated by

a. $\dfrac{x_1 + x_2}{2}$

b. $\dfrac{x_1 - x_2}{\ln(x_1/x_2)}$

c. $\sqrt{x_1 x_2}$

or

d. $\dfrac{2}{\dfrac{1}{x_1} + \dfrac{1}{x_2}}$?

One final important point should be mentioned at this point: all of the mean measures discussed so far have been defined on finite data sets. If, on the other hand, the sets are extremely large, for instance, containing several hundred million elements or perhaps even an infinite number, alternative definitions for AM, GM, HM need be formulated. To this end you must resort to the integral calculus as follows. If the data values are represented by a function, f, defined over an interval, [a,b], the arithmetic mean of the values, denoted by AM(f), is defined as

$$AM(f) = \frac{1}{b-a} \int_a^b f(x)dx.$$

As for the geometric mean, since

$$GM = (x_1 x_2 \cdots x_n)^{1/n} = e^{\Sigma \, (1/n)\ln(x_i)}$$

if the function f is always positive, the geometric mean of f, GM(f), is defined as

$$GM(f) = e^{\frac{1}{b-a} \int_a^b \ln(f(x))dx}$$

Similarly, the harmonic mean of f, where $f > 0$, is given by

$$HM(f) = \frac{b - a}{\int_a^b \dfrac{1}{f(x)} \, dx} .$$

Suppose that the data set consists of the 10 scores $e^{.1}$, $e^{.2}$, $e^{.3}$, $\cdots$, $e^{1.0}$. In this case the approximate values of the three means are AM = 1.81, GM = 1.73 and HM = 1.66. If, on the other hand, the data set consists of the 100,000 scores $e^{.00001}$, $e^{.00002}$, $\cdots$, $e^{1.0000}$, the three means can best be approximated by their integral definitions:

$$AM = \frac{1}{1} \int_0^1 e^x dx = e-1 = 1.72$$

$$GM = e^{\int_0^1 \ln(e^x)dx} = e^{1/2} = 1.65$$

$$HM = \cfrac{1}{\int_0^1 e^{-x}dx} = \frac{e}{e-1} = 1.58.$$

EXERCISES

1. Write a program that serves a dual purpose. First, have it compute the arithmetic, geometric, and harmonic mean for input of $n \le 100$ positive numbers. Be careful with the geometric mean: from the standpoint of memory capacity and computer roundoff error, it might be best to use the formula

$$GM = e^{\frac{1}{n} \Sigma \ln(x_i)}$$

Secondly, if x_1 and x_2 are two positive numbers, and if $AM(x_1,x_2)$, $LM(x_1,x_2)$, $GM(x_1,x_2)$ and $HM(x_1,x_2)$ are the four associated mean values, have your program compute

   a. $AM(\ AM(x_1,x_2),\ HM(x_1,x_2)\ )$
   b. $LM(\ AM(x_1,x_2),\ HM(x_1,x_2)\ )$
   c. $GM(\ AM(x_1,x_2),\ HM(x_1,x_2)\ )$
   d. $HM(\ AM(x_1,x_2),\ HM(x_1,x_2)\ )$

   and compare each of these values with $GM(x_1,x_2)$ and $LM(x_1,x_2)$ to see which is closest. Is there a pattern?

2. For the two sets of positive data $\{x_1,x_2,\bullet\bullet\bullet,x_n\}$ and $\{y_1,y_2,\bullet\bullet\bullet,y_n\}$, it is a property of the geometric mean that

$$\sqrt[n]{(x_1+y_1)(x_2+y_2)\bullet\bullet\bullet(x_n+y_n)} = \sqrt[n]{x_1 x_2 \bullet\bullet\bullet x_n} + \sqrt[n]{y_1 y_2 \bullet\bullet\bullet y_n}.$$

Prove this in the special case when n = 2.

BUFFON'S NEEDLE PROBLEM

Much of the classical textbook material on introductory probability focuses on rather simple theoretical probabilities and on the laws that govern these probabilities when certain initial conditions are met. These laws might include whether to multiply or add the associated probabilities, $P(A \cap B) = P(A) \cdot P(B)$, $P(A \cup B) = P(A) + P(B)$, depending on whether the events are independent or mutually exclusive, or possibly divide a couple of probabilities, as prescribed by a conditional probability.

Now it is certainly true that these mathematical relationships hold true for all kinds of probabilities, but it is necessary to distinguish between the theoretical and the subjective. The former could be illustrated by the probability of drawing a jack from a full shuffled deck of cards, while the probability that Jack Nicklaus will win next year's Masters typifies the latter. The probability of drawing a jack is clearly 4/52, but who can say what the probability of Nicklaus winning the tournament is. Subjective probabilities just can't be given a definite value: they can only be approximated because there is no way of ascertaining the true value. There is no way of conducting an experiment or of running a series of trials to get an approximate value. In contrast to this are the events whose probabilities, if by chance they can't be computed as quickly as the probability of drawing the jack, can be sufficiently approximated by repeated independent trials. One example of this is to flip a thumb tack, and inquire about the probability that the tack will land point-up. If, after 500 flips, the point lands up a total of 373 times, you can say that the desired probability is approximately .746, and this is termed the empirical probability. But how can the true probability be determined? It's quite likely that it can't. One redeeming point is that the Law of Large Numbers furnishes us [41] with the guarantee that as the number of trials increases without bound, the empirical probability approaches the theoretical probability. Usually it is just a matter of how many trials are sufficient to get a good enough approximation. The setting is now ready for the main topic of this section, Buffon's Needle Problem.

Buffon was an 18th century French naturalist who went by the name Count Buffon (though officially he was George Louis Leclerc). His name is passed along in the science circles solely because of his needle problem, which he posed as follows.

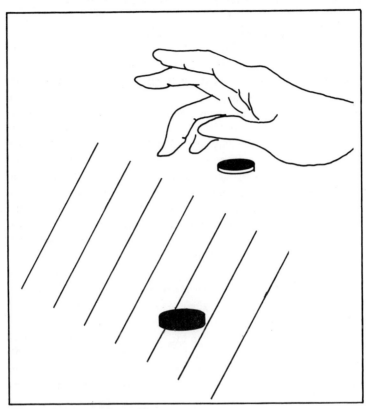

Fig. 6-3. Buffon's needle problem.

A needle of length L is dropped from a large enough height onto a plane surface ruled by parallel lines (Fig. 6-3). The distance between successive lines is of common value D. What is the probability that the needle will intersect one of the parallel lines?

There is no doubt that this is a simple enough looking problem; in fact, so simple and unimportant that you should wonder why it wasn't lost in Buffon's memoirs. Fortunately, it does have a rather attractive solution, and it is quite instructive to those studying more advanced probability theory—more will be said along these lines later. For the time being, see if you can't write a program that will simulate the random dropping of a needle onto a ruled surface. One approach is as follows.

The ruled surface will be the x-y coordinate system with the lines $x = 0$, $x = \pm 1$, $x = \pm 2$, ••• serving as the parallel lines whose common distance apart is $D = 1$. The left end of the needle (say the pointed end) will be randomly selected by calling the random number generator function RND in BASIC. The needle end will then correspond to some point (x_1, y_1) in the plane. But since the random numbers so generated are given as decimals with values between 0 and 1, the left end of the needle will be located somewhere in the unit square $0 \leq x \leq 1$, $0 \leq y \leq 1$. Even though this unit square as shown in Fig. 6-4 is the only portion of the entire plane that the left end of the needle will land in, it serves the same purpose as if the entire plane were used. The location of the right end of the needle can be computed, in one way, by specifying the angle α determined by the needle and the positive x-axis. This angle, which must be between 0 and 2π radians, can be found by again calling RND and setting $\alpha = 2\pi \cdot$RND. Since the needle is of length L, the right end, given by the point (x_2, y_2), is located at

$$x_2 = x_1 + L \cdot \cos(\alpha)$$
$$y_2 = y_1 + L \cdot \sin(\alpha).$$

This gives the diagram shown in Fig. 6-5. It follows that the needle will intersect one of the parallel lines if and only if $|x_1 - x_2| \geq 1$. An equivalent way of stating this is to say that the needle will intersect one of the parallel lines if $[x_1] \neq [x_2]$, where the brackets denote the greatest integer function.

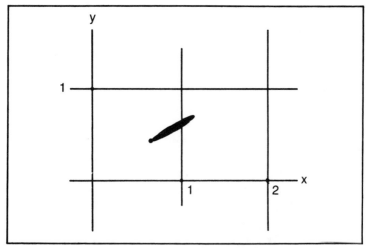

Fig. 6-4. The left end of needle falls in a unit square.

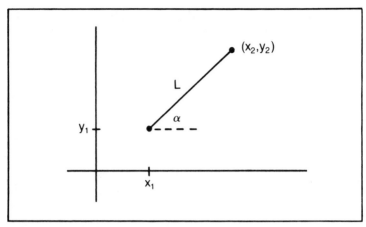

Fig. 6-5. The right end of the needle.

Below is a program that allows you to drop the needle 500 times and record the number of line crossings (given by NCROSS).

Program

```
110   REM BUFFONS NEEDLE PROBLEM
120   INPUT L
130   NDROPS = 0
140   NCROSS = 0
150   REM FIRST WE PICK A POINT(X,Y) IN UNIT
      SQUARE AT RANDOM
160   X1 = RND
170   Y1 = RND
180   REM THE POINT (X1,Y1) IS THE LEFT END OF
      THE NEEDLE
190   ALPHA = 2+PI*RND
200   X2 = X1 + L*COS(ALPHA)
210   Y2 = Y1 + L*SIN(ALPHA)
220   REM THE RIGHT END OF THE NEEDLE IS (X2,Y2)
230   IF INT(X1) = INT(X2) GO TO 250
240   NCROSS = NCROSS + 1
250   NDROPS = NDROPS + 1
260   IF NDROPS = 500 GO TO 280
270   GO TO 160
280   RATIO = NCROSS/NDROPS
290   PRINT "WHEN THE NEEDLE LENGTH IS"; L
```

300 PRINT "THE EMPIRICAL PROBABILITY FOR BUF-
    FON IS"; RATIO
310 STOP
320 END

Sample results from running this program include the following.

Length	.5	.8	1	2	3	4
Empirical Probability	.32	.51	.632	.838	.9	.926

It is fairly obvious that as L decreases to 0 so does the likelihood that the needle crosses a line; while as L increases, the likelihood goes towards one.

What makes Buffon's Needle Problem so attractive is that in the special case when the needle length equals the spacing between the parallel lines, $L = D$, an exact figure for the probability of a crossing has been determined. This probability is $2/\pi$. How amazing it is to see this irrational and transcendental number π appear in the answer; a number that is defined in terms of the circumference of a circle and that often appears in closed form evaluating some infinite mathematical series or a definite integral. This result is somewhat surprising, not only because the probability is expressed in such nice form, but because the grid lines are all straight lines, not curves, and customarily such a situation would connote a numerical value that is rational.

The decimal value of $2/\pi$ is approximately .6366, so you can see from the above listing that the empirical probability of .632 was quite close. Interestingly enough, many experiments have been made on this since Buffon first posed the problem. One such experiment [42] was performed in 1901 by an Italian mathematician named Lazzerini, who obtained an empirical probability of .63661972 after 3408 drops. This figure, when divided by 2, gives an approximate for π that is accurate to within 3×10^{-7}.

You might be sufficiently curious by now as to want to investigate how the value $2/\pi$ was obtained or to inquire what the probability is when $L \neq D$. As for the latter, you may wish to consult [65], which furnishes two solutions, depending on whether $L \leq D$ or $D \leq L$. The two references [65] and [43] give different approaches to evaluating the probability P when $L = D$. Both discussions require some integral calculus background because in [65] the value of P is given by the integral

$$P = \int_0^\pi \frac{L\sin(\theta)}{\pi D} \, d\theta$$

while in [43] we have a double integral expression,

$$P = \frac{\displaystyle\int_0^{\pi/2} \int_0^{L\cos(\theta)/2} dx d\theta}{\displaystyle\int_0^{\pi/2} \int_0^{D/2} dx d\theta} \, .$$

Fortunately, the integrals simplify quite nicely—again proving there is mathematical justice in the world!

A rather interesting real life variation of Buffon's Needle Problem is discussed in [24]. This problem concerns the probability of a small larval fish being killed by an impellar blade of an intake pump as water is being used to cool the reactor in a nuclear power plant.

Exercises

1. Write a program, different from the one given, which simulates the random drop of a needle of length L onto a ruled surface of common spacing D. Test your program for $L = D$ and see if you get an empirical probability of approximately $2/\pi$. Also test the case $2L = D$ and see if the probability is close to $1/\pi$. For the situation $L \geq D$ it is interesting not only to note the likelihood of crossing a line, but also how many lines are crossed. Consequently, set $L = kD$ ($k = 1,2,\cdots,10$) and obtain empirical data for the expected number E of lines that are crossed when a needle of length L is dropped on the ruled surface.
2. A circular coin, of radius 1, is dropped at random onto a ruled surface, with lines spaced apart at a common distance of 2. What is the probability that the coin will intersect one of the lines?

THE LINE OF BEST-FIT

The primary purpose of linear correlation analysis is to measure the strength of a linear relationship between two variables in question. In business, industry, government, education, agriculture, and in many other fields, two variables under consideration may bear a strong mathematical relationship to one another. This relationship could be any one of many possibilities but the more common ones are the exponential, logarithmic, quadratic and linear.

To illustrate the former, a biologist might be studying the bacteria count in a particular culture under a prescribed set of environmental conditions. The two variables involved are time (t) and number (N) of bacteria present at time t. Some sample data might include that presented in Table 6-6. If this data is sketched as a graph as shown in Fig. 6-6, you get what appears to closely follow an exponential growth as described roughly by the equation $N = 10000(2)^{t/2}$.

t	N
0	10000
1	14000
2	20000
3	29000
4	40000
5	55000
6	75000

Table 6-6. A Bacteria Count.

Linear relationships occur only as frequently as exponentials, although they are discussed and studied much more in standard courses. Typically, you might be interested in the expected weight (in pounds) of an adult male between, say, the ages of 25 and 45. A survey of men with ages 25, 27, 29, •••, 45 could yield the results shown in Table 6-7.

Looking at the associated scatter diagram in Fig. 6-7 you can readily observe a high degree of linear correlation between the variables age (t) and weight (w).

Computation of Pearson's product moment r (also known as the coefficient of linear correlation [41]) produces a value of $r = .998$:

$$r = \frac{11(72072) - 385(2034)}{\sqrt{11(13915) - 385^2} \sqrt{11(377880) - 2034^2}}$$

$$= .998$$

334

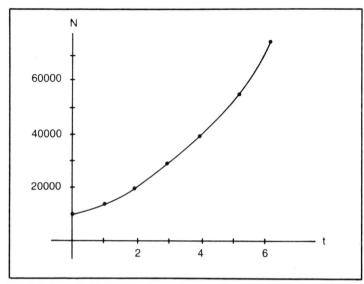

Fig. 6-6. An exponential curve approximating the bacteria count.

which, since it is close to one, indicates an extremely high degree of linear correlation. You should always be careful in interpreting the true meaning of linear correlation. It doesn't imply that there is a cause-and-effect relationship between the two variables. Rather the data indicates there is a mathematical relationship between them.

The line of best-fit for the above age-weight example is computed to be $w = 2.005t + 114.752$ which, for all practical purposes, can be replaced by the simpler equation $w = 2t + 115$.

Table 6-7. A Chart of Male Weights.

Age	Weight
25	164
27	170
29	173
31	178
33	180
35	185
37	188
39	193
41	196
43	201
45	206

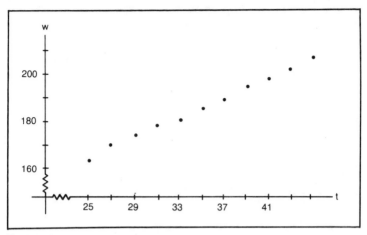

Fig. 6-7. Graphing male weight vs age.

The real function of the line of best-fit, in general, lies in predicting values of one of the variables, usually the dependent variable, given the value of the other. Thus, it might be reasonable to say that the average weight of a male of age 36 is $2(36) + 115 = 187$ pounds. Another word of caution though: it might be the situation that the line of best-fit, say $y = mx + b$, is not very useful in predicting dependent variable scores. It might not be any more useful than the constant curve $y = \overline{y}$, and this is precisely where a regression analysis test (H_o, the line of best-fit, is of no value in predicting scores) would be employed.

Because of its importance, the line of best-fit deserves a little more discussion. Assuming our data set includes the points (x_1, y_1), $(x_2, y_2), \bullet\bullet\bullet, (x_n, y_n)$, the line of best-fit $y = mx + b$ is that unique line that somehow comes the closest to all the n points in the data set. Since the words "comes the closest" are ambiguous, a more precise wording is needed. Consequently the line of best-fit is defined as that line which minimizes the sum

$$\sum_{i=1}^{n} |y_i - (mx_i + b)|.$$

This sum is represented graphically in Fig. 6-8 as the total of the lengths (all lengths are considered nonnegative) of the n vertical distances as shown.

336

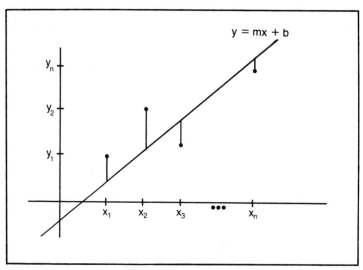

Fig. 6-8. The line of best-fit.

Mathematically, you can use some of the basics from introductory differential calculus to derive the equation of the line of best-fit, by showing that the two unknown parameters m and b must have the value [70] given by

$$m = \frac{n \sum x_i y_i - \sum x_i \sum y_i}{n \sum x_i^2 - (\sum x_i)^2}$$

$$b = \frac{\sum y_i \sum x_i^2 - \sum x_i \sum x_i y_i}{n \sum x_i^2 - (\sum x_i)^2} \cdot$$

An interesting geometric approach for determining these values can be found in [77].

In an effort to alter the concept of the line of best-fit so that it still has some meaning, one interesting alternative would be to minimize the sum of the distances that the points (x_i, y_i) are away from the line itself. This means that you should minimize the sum of the distances d_i, where the d_i (again all nonnegative) are the perpendicular distances between the points (x_i, y_i) and the line as shown in Fig. 6-9. Even though this line may not have much statistical meaning, such as not being useful in prediction, it does carry the above mathematical meaning. So as to avoid confusion, this line will

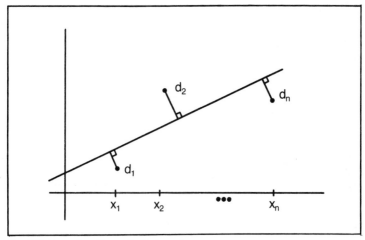

Fig. 6-9. Line of mathematically-best-fit.

be called *the line of mathematically-best-fit,* abbreviated by LOMBF, as opposed to the initial line of best-fit, LOBF. The student who has been exposed to the fundamentals of partial differentiation ought to be able to understand the following sequence of steps leading to the equation of the line LOMBF.

First, you must recall from calculus that the distance between the point (p,q) and the line $Ax + By + C = 0$ is given by the quotient

$$\frac{|Ap + Bq + C|}{\sqrt{A^2 + B^2}}.$$

Consequently, if the line LOMBF is expressed in the form $y = mx + b$, each distance d_i from above is equal to the quotient

$$d_i = \frac{|mx_i - y_i + b|}{\sqrt{m^2 + 1}}.$$

In order to minimize the sum of the d_i, you would have to cope with the absolute value function, and it is somewhat annoying to keep track of whether the derivative of this function is positive or negative, or maybe even fails to exist. To circumvent this problem, you shall attempt to minimize the sum of d_i^2 (hence the phrase least-squares), which is just as good a problem because it gives the same solution set. To this end, note that

338

$$\sum d_i^2$$

is actually a function, say f(m,b), of the two variables m and b, given by

$$f(m,b) \;=\; \sum_{i=1}^{n} \frac{(mx_i - y_i + b)^2}{m^2 + 1}.$$

The technique used in minimizing f is to solve the system of two equations $f_m = 0$ and $f_b = 0$, where f_m and f_b represent the two partial derivatives of f. These yield the following equations, respectively:

$$0 = \Sigma \, (mx_i - y_i + b) \, (x_i + my_i - mb)$$
$$0 = \Sigma \, (mx_i - y_i + b).$$

Solving for b in the second equation gives the quotient

$$b \;=\; \frac{\Sigma \, y_i - m \, \Sigma \, x_i}{n}.$$

Substituting this into the first equation produces a complicated quadratic equation in m, which when simplified, reduces to $m^2 + Dm + 1 = 0$, with D having the value

$$D \;=\; \frac{n(\Sigma \, x_i^2 - \Sigma \, y_i^2) + (\Sigma \, y_i)^2 - (\Sigma \, x_i)^2}{n \, \Sigma \, x_i y_i - \Sigma \, x_i \, \Sigma \, y_i}.$$

Since the quadratic could have two different solutions, care must be taken to select the correct value for m. This can usually be ascertained by checking for the proper algebraic sign of m.

To illustrate the above, consider the six piece data set of (1,1), (2,1), (3,2), (4,2), (5,3) and (6,5). The value of D is

$$D \;=\; \frac{6(91 - 44) + 14^2 - 21^2}{6(62) - 21(14)} \;=\; \frac{37}{78} \;=\; .4743.$$

Thus the equation $m^2 + .4743m + 1 = 0$ has the two zeros, .79056 and -1.26492. Only the first of these makes sense. The value of the y-intercept is then given by

$$b = \frac{14 - .79056(21)}{6} = -.4336.$$

The line of mathematically-best-fit is approximately $y = .79056x - .4336$. The corresponding values of d_i are $d_1 = .58997$, $d_2 = .13534$, $d_3 = .05681$, $d_4 = .66850$, $d_5 = .47635$ and $d_6 = .63327$. This implies the minimum value of the sum of the d_i is approximately 2.56024, and that the function value of f is

$$f(.79056, -.4336) = 1.44444.$$

Exercises

1. Write a program which computes the equation of the line of mathematically-best-fit for a general data set (x_1, y_1), (x_2, y_2), $\cdots$, (x_n, y_n). Compute the value of each d_i, and print out the value of $f(m,b)$. Then try to find an arrangement of data points for which $f(m,b) = 1$.

2. The expression for D is a rational expression, with a denominator of

$$n \, \Sigma \, x_i y_i - \Sigma \, x_i \, \Sigma \, y_i.$$

Explain the geometric significance when this denominator equals zero.

THE QUADRATIC OF BEST-FIT

As mentioned in the previous section, linear correlation analysis is the study of the linear relationship between two variables. In fact, many variables do obey a linear relationship, so it is important to study linear correlation and regression. It only seems natural that after exhausting the study of linear analysis, you turn to quadratic analysis: the study of the quadratic relationship between two variables.

Some of the world's great formulas involve the quadratic tie between two extremely important and very pertinent variables. To illustrate, a projectile fired upward with an initial velocity v_0 attains the approximate height h after t seconds of $h = -16t^2 + v_0 t$, and the everlasting gem of geometry, the formula for the area A of a circle of radius r is $A = \pi r^2$. A more complex example found in the chemistry realm asserts that the maximum end-to-end distance between two methyl groups in the molecule n-butane is (under certain conditions imposed on the bond angles and the rotations about the bonds) given by $d = 4.56L^2$, where L is the length of a carbon-carbon link [18]. Many relationships in the real world are quadratic in nature, and the variables may represent quantities from any field conceivable.

It is important to recognize which situations might lead to a quadratic connection. For instance, suppose the variables are p and q, and suppose the data set consists of the five pairs (p,q) of (1,0), (2,8), (3,14), (4,16) and (5,17). When this data is graphically presented on a p-q coordinate system, it appears that the points lie sufficiently close to the graph of a parabola as shown in Fig. 6-10 that it would be appropriate to find and use the quadratic relationship (q $= Ap^2 + Bp + C$) as determined by the above data. But, those who are familiar with many of the common elementary mathematical functions may note that these points could be suitably approximated by a logarithmic curve, say $q = A\log(p)$, as shown in Fig. 6-11.

The point here is that there could be many curves (and, in fact, there are infinitely many polynomials [63]) that pass through a prescribed set of N points. Consequently, the situation might help dictate which curve is appropriate to approximate the points. If you know that eventually an increase in p causes a decrease in q, it would not be suitable to use the logarithmic curve. If an increase in p

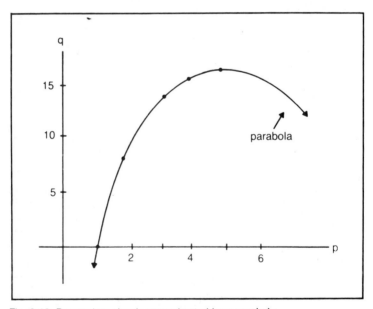

Fig. 6-10. Data points closely approximated by a parabola.

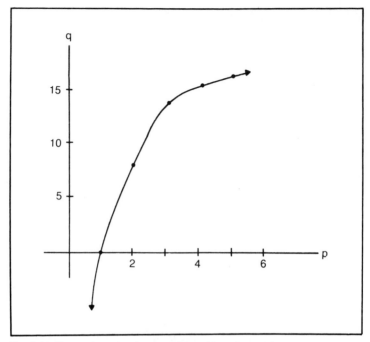

Fig. 6-11. Data points closely approximated by a log curve.

always causes q to increase, the quadratic model might be discarded although a portion of the curve could prove useful.

Assume for the present that a quadratic relationship is anticipated between two variables. The next step is to try to determine, as best as you can, the extent of this quadratic form. In other words, you need to determine the quadratic equation that best describes the interaction between the variables. To this end, proceed as you did in the previous section when you determined the line of best-fit, LOBF.

First, suppose your data set consists of the N points (x_1, y_1), (x_2, y_2), $\bullet\bullet\bullet$, (x_N, y_N), and you wish to find the quadratic $y = Ax^2 + Bx + C$ that best fits this data set as shown in Fig. 6-12. More precisely, you need to find the three coefficients A, B, and C so that the least-squares sum

$$\sum_{i=1}^{N} d_i^2$$

342

is minimized. As in the previous section, the expression d_i represents the vertical distance between the curve and the point (x_i, y_i). The value of d_i is given by $d_i = |y_i - (Ax_i^2 + Bx_i + C)|$. The sum of squares of d_i can be viewed as a function f of three variables A, B, and C where

$$f(A,B,C) = \sum_{i=1}^{N} (y_i - (Ax_i^2 + Bx_i + C))^2.$$

In order to minimize f, compute the three partial derivatives f_A, f_B, and f_C, set each to zero, and solve. Thus you have

$$f_A = 0 \text{ implies } \sum_{i=1}^{N} 2 \, [y_i - (Ax_i^2 + Bx_i + C)] \, (-x_i^2) = 0$$

$$f_B = 0 \text{ implies } \sum_{i=1}^{N} 2 \, [y_i - (Ax_i^2 + Bx_i + C)] \, (-x_i) = 0$$

$$f_C = 0 \text{ implies } \sum_{i=1}^{N} 2 \, [y_i - (Ax_i^2 + Bx_i + C)] \, (-1) = 0.$$

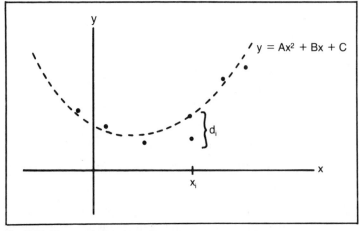

Fig. 6-12. The quadratic of best-fit.

Solving for C from this last equation gives

$$C = \frac{\Sigma\, y_i - A\, \Sigma\, x_i^2 - B\, \Sigma\, x_i}{N}$$

and substituting this into the middle equation yields

$$B = \frac{\Sigma\, x_i\, \Sigma\, y_i - N\, \Sigma\, x_i y_i - A\, \{\, \Sigma\, x_i\, \Sigma\, x_i^2 - N\, \Sigma\, x_i^3\,\}}{(\Sigma\, x_i)^2 - N\, \Sigma\, x_i^2}\, .$$

For the sake of brevity let the above denominator be denoted by D, so

$$D = (\Sigma\, x_i)^2 - N\, \Sigma\, x_i^2\, .$$

Substituting the expressions for C and B from above into the first equation produces the simple linear equation $Au + v = 0$ where u and v are complicated expressions in x_i and y_i, namely

$$u = \frac{(\Sigma\, x_i\, \Sigma\, x_i^2 - N\, \Sigma\, x_i^3)(\Sigma\, x_i^3 - (1/N)\, \Sigma\, x_i\, \Sigma\, x_i^2)}{D}$$

$$- \Sigma\, x_i^4 + \frac{1}{N}\, (\Sigma\, x_i^2)^2$$

$$v = \frac{(\Sigma\, x_i\, \Sigma\, y_i - N\, \Sigma\, x_i y_i)((1/N)\Sigma\, x_i\, \Sigma\, x_i^2 - \Sigma\, x_i^3)}{D}$$

$$+ \Sigma\, x_i^2 y_i - \frac{1}{N}\, \Sigma\, x_i^2\, \Sigma\, y_i\, .$$

Thus, the value for A is $-v/u$, and then this value can be used in the second equation to obtain B, and both of these values can be

344

substituted in the C = equation to obtain C. From the programming standpoint, there are several instances here where the division process may yield either a denominator of zero or a denominator that is almost zero—so beware!

To illustrate how this method works, use the five data points (1,0), (2,8), (3,14), (4,16), (5,17) as mentioned earlier and compute the following:

$$\Sigma x_i = 15 \qquad \Sigma x_i^2 = 55$$
$$\Sigma y_i = 55 \qquad \Sigma x_i^2 y_i = 839$$
$$\Sigma x_i y_i = 207 \qquad \Sigma x_i^3 = 225$$
$$\Sigma x_i^4 = 979.$$

You then get the values $D = -50$, $u = -14$, $v = -18$, $A = -1.2857$, $B = 11.9142$ and $C = -10.5999$. Thus, the quadratic curve of best fit is given by the equation $y = -1.28x^2 + 11.91x - 10.60$ as shown in Fig. 6-13. This is actually a fairly good fit because the five differences d_i are $d_1 = .03$, $d_2 = .10$, $d_3 = -.40$, $d_4 = .56$ and $d_5 = -.05$, and they are all quite small.

Finally, it should be mentioned that an important rationale for determining the quadratic approximation to data scores, just as with the linear approximation, lies in the category of prediction. Statisticians need to be able to predict values of the dependent variable

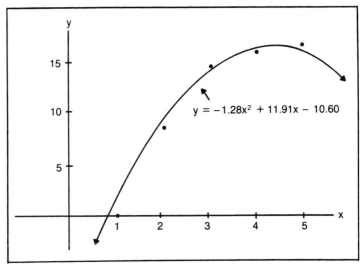

Fig. 6-13. Quadratic best-fit of some data points.

given some value for the independent variable. So, for example, using the data set above, if x were given the value of 2.5, it would make sense to say the corresponding y value would be 11.175 (i.e., $-1.28(2.5)^2 + 11.91(2.5) - 10.60$).

Exercises

1. Write a program that computes the equation of the quadratic of best fit for a general data set (x_1, y_1), $\bullet\bullet\bullet$, (x_N, y_N). Print out the data set, the quadratic equation, and the individual values of d_i. Test your program with the data from this section.
2. Consider the data set (1,6), (2,11) and (3,18). These points happen to lie on the parabola $y = x^2 + 2x + 3$. Using the equations derived above for the quadratic of best fit, what do you think will be the values of A, B, and C? What do you think will happen if you only consider the two-point data set (1,6), (2,11)?

CHEBYSHEV'S INEQUALITY

Suppose that Professor Jones returns 30 graded exam papers to his statistics class. He informs the class that the mean test score was 78. Student Sammy scored an 83 on his paper, and promptly reasons that since his score was slightly above average, his grade would be in the C+ to B− range.

Unfortunately, Sammy is guilty of faulty reasoning: it's faulty because he has made an assertion about the relative standing of a particular score with only the knowledge of one measure (the mean) of central tendency to base his conclusion on. If Sammy also knew the values of the mode or the median, he could better assess how his score stacks up against his classmates. And, better yet, if Sammy knew how the test scores were distributed, how they were dispersed, he could better describe the relative merits of any individual score.

One of the measures of data dispersion is the standard deviation. If the standard deviation is small, then the scores all pretty much tend to be grouped close together, around the mean. Thus, if Professor Jones declares that the standard deviation of the scores is 1.5, Sammy should change his mind about his possible grade score, and perhaps raise it to an A. If the standard deviation is large, the scores are widely dispersed. If Professor Jones declares that the standard deviation is 8, then Sammy would know there are probably quite a few scores higher than his 83. His chances of getting even a B grade would be minimal.

The problem that naturally arises is whether a particular value for the standard deviation is small or large. These words carry different meanings for different people. So even though the standard deviation gives a measure of the fluctuation of the data, its true interpretation is still not clear. One means for helping to understand and clear-up this interpretation is furnished by the following inequality, attributed to the Russian mathematician Chebyshev.

Chebyshev's Inequality: Suppose that X is a random variable (discrete or continuous) having mean μ and standard deviation σ, which are both finite. Then, if ϵ is any positive number, the probability that X differs from the mean by at least ϵ is at most equal to σ^2/ϵ^2. Symbolically this is written,

$$P(\mid X - \mu \mid \geq \epsilon) \leq \sigma^2/\epsilon^2.$$

To give a simple illustration, consider the variable X that assumes the values 1,2,3 with equal likelihood. The mean will be $\mu = 2$. Suppose you set $\epsilon = 1$. Then the probability that X differs from 2 by at least 1 ($P(\mid X - 2 \mid \geq 1)$), is the same as the probability that X equals either 3 or 1, which is $P = 2/3$. Chebyshev's inequality says this probability is at most $\sigma^2/1^2 = \sigma^2$, and the value of σ^2 happens to be 2/3 as shown in Table 6-8. Likewise, the probability that X differs from 2 by at least $\epsilon = .9$ is again 2/3 (because of X = 1 or X = 3), while Chebyshev's inequality gives this probability a maximum value of $\sigma^2/\epsilon^2 = .82$.

What is remarkable about this result is that a statement is made about the distribution of a random variable, a statement that incorporates the mean and standard deviation, but which is devoid of the specific probability distribution of X. A formal proof of Chebyshev's Inequality can be found in [12].

Since the value of ϵ can assume any positive number, you can make the substitution $\epsilon = k\sigma$ with k being any positive number (though it is commonly an integer). Then the inequality becomes

$$P(\mid X - \mu \mid \geq k\,\sigma) \leq \frac{1}{k^2}$$

which can be interpreted as, "the probability is at most $1/k^2$ that X will differ from the mean by at least k standard deviations." Con-

Table 6-8. Computing a Specific Standard Deviation.

x	P(x)	xP(x)	x²P(x)
1	1/3	1/3	1/3
2	1/3	2/3	4/3
3	1/3	3/3	9/3
		6/3	14/3

$$\sigma^2 = \Sigma \ x^2P(x) - [\Sigma \ xP(x)]^2$$
$$= 14/3 - 4$$
$$= 2/3.$$

sequently, the proportion of scores that are at least k standard deviations from the mean is at most $1/k^2$. For example, with $k = 2$, you know that at most 25 percent $(1/4 = .25)$ of the scores are two or more standard deviations away from the mean as shown in Fig. 6-14. Equivalently, at least 75 percent of the scores lie within 2 standard deviations of the mean! Keep in mind that these results apply to any distribution as long as the mean and variance are finite.

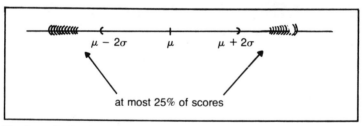

Fig. 6-14. Chebyshev's Inequality with $k = 2$.

For a more involved set of data that illustrate Chebyshev's Inequality, consider the frequency distribution listed in Table 6-9. The mean is given by

$$\mu = (1/n) \ \Sigma \ xf = 5.11$$

and standard deviation

$$\sigma = \sqrt{\frac{n \ \Sigma \ x^2 \ f - (\ \Sigma \ xf)^2}{n(n-1)}} = 1.895.$$

x	f
1	2
2	4
3	13
4	18
5	27
6	17
7	9
8	3
9	4
10	3
	100

Table 6-9. A Sample Frequency Distribution.

According to Chebyshev's Inequality for $k = 2$, at most 25 percent of the scores are either greater than $5.11 + 2(1.895) = 8.9$, or less than $5.11 - 2(1.895) = 1.32$. Since there are actually nine scores in this range (two 1s, four 9s, three 10s) the exact proportion is $9/100 = 9$ percent.

To write a program that will verify Chebyshev's Inequality for various values of k and for an arbitrary set of data, assume that N data scores $x_1, x_2, \cdots, x_N$ are to be fed into the program. N is also input. You might as well assume all the scores are positive. The mean μ and standard deviation σ will need to be computed, where

$$\mu = (1/n) \Sigma x$$

$$\sigma = \left[\frac{N \ \Sigma \ x^2 - (\Sigma x)^2}{N(N-1)} \right]^{1/2}.$$

An arbitrary value for $k > 1$ will be given. You should then count the number of data scores x_i that are either greater than $\mu + k\sigma$ or less than $\mu - k\sigma$. This proportion should be less than $1/k^2$.

Exercises

1. Write the program as described above. Test it with the specific values $k = 1.5$, $N = 25$, and the scores

    1,2,3,4,5,6,7,8,9,10,3,4,5,6,10,3,4,5,6,5,6,5,5,5,5.

Print out the scores, μ, σ, the number and proportion of scores that are at least k standard deviations away from the mean, and the proportion indicated by Chebyshev's Inequality.

2. A random variable X has a mean of 3 and a variance of 2. Use Chebyshev's Inequality to obtain an upper bound (maximum value) for the probability

$$P(\mid X - 3 \mid \geq 2).$$

BINOMIAL DISTRIBUTION

There are a myriad of probability distributions—more than you could ever hope to study. Fortunately, though, most of these distributions occur in such rare cases that it is impractical to investigate them to any depth. The remaining limited number of types, or families, of distributions that are used in a wide range of applications include the following frequently studied distributions:

    a. the uniform distribution
    b. the normal distribution (the Gaussian distribution)
    c. the Poisson distribution
    d. the exponential distribution
    e. the chi-square distribution
    f. Student's t distribution
    g. the F distribution
    h. the binomial distribution.

The binomial distribution will be concentrated on for the remainder of the section. To begin with, a Bernoulli (named after James Bernoulli, who investigated these matters at the end of the 17th century) trial is a random trial that has precisely two possible outcomes, such as sink or swim, heads or tails, or success or failure. These two outcomes are usually quantified by assigning the values 0 and 1 to them. Which outcome is assigned the value 0 and which is assigned the value 1 is arbitrary. In electronics, quite often the value 1 will indicate that a particular network junction has current flowing through it, while 0 would indicate that the current is off. Computer hardware used to read input data on punched cards by determining whether or not a beam of light passed through a punched hole. If there was no hole punched, the computer would assign a 0 to that particular bit location; otherwise the value 1 would be assigned.

Bernoulli trials are really quite common and occur often in

everyday life. It is customary to label the two possible outcomes as success and failure, to let p denote the probability of the successful outcome, and to let q (where q = 1 − p) denote the probability of failure.

In a game of poker, if the successful event is that of drawing an ace from a full shuffled deck, then p = 4/52 and q = 48/52. If, on the other hand, you are interested in drawing a black face card (jack, queen, king of Clubs or Spades) then p = 6/52 and q = 46/52.

Suppose Joe Tourist is spending the weekend in Las Vegas, and he decides to try his luck at the tables, notably roulette. For those unfamiliar with the game, a roulette wheel is usually divided into 38 equal slots, numbered 0,00,1,2,3, •••, 36. The slots numbered 0 and 00 are colored green, and the slots numbered 1,3,5, •••, 35 are colored black, while the even numbered slots are colored red as shown in Fig. 6-15. A ball is projected along a path that borders the rim of the wheel, and when its velocity decreases to a certain point, the ball falls into one of the 38 slots. There are a variety of ways you can place a bet, but assume that Joe bets that the ball will land in a red slot (either the 2,4,6, •••, 36 numbered slot). Since there are 18 red slots, the probability of success p for Joe is therefore 18/38. Joe has almost an even chance of winning. The payoff for this bet is 1-1, which means that your net profit is the

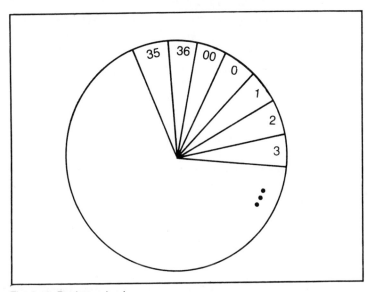

Fig. 6-15. Roulette wheel.

same as the amount of your bet. If you bet $5 and win, the house will return $10 to you, giving you a $5 profit.

As important as the individual Bernoulli trials are, people are usually more interested in a sequence of Bernoulli trials. To this end, suppose you have n Bernoulli trials; the trials are all statistically independent (what happens in one trial has no bearing on the outcome of any other trial); and the probability of success for each trial is the common value p. If these conditions are met, you have what is called a Bernoulli process. If you draw a single card from a full shuffled deck in hopes of getting an ace you do this a total of 5 times, each time replacing the drawn card back into the deck, this is a Bernoulli process. You should note the importance here of replacing the card. If Joe Tourist decides to stay at the roulette wheel for a while and continue to place the same kind of bet, always betting on red, this is also a Bernoulli process. This follows because the ball either lands in red (p = 18/38) or it doesn't (q = 20/38), and each spin is independent of the previous spins (assuming the wheel isn't crooked).

Our chief interest in the Bernoulli process is in how many times our successful event will occur, and what the probability is that it will occur a given number of times. If the roulette wheel is spun five times, how many times will the ball land in a red slot? What is the probability that the ball will land in the red slot three times? To answer this latter question refer to the Bernoulli probability function, f, which is better known as the binomial probability function. Thus, if p denotes the probability that an event will happen in any single Bernoulli trial (called the probability of success), and if q = 1 − p, then the probability that the event will happen exactly i times in a sequence of n trials is given by the probability function

$$f(i) = \text{Prob}(X = i) = \binom{n}{i} p^i q^{n-i}$$

where the random variable X denotes the number of successes in n trials, and i = 0, 1, 2, •••, n. The symbol $\binom{n}{i}$ is the well-known binomial coefficient, defined by

$$\binom{n}{i} = \frac{n!}{i!\,(n-i)!} \;.$$

Return to our friend, Joe Tourist, at the roulette table: if Joe bets on red for five consecutive spins, the probability that he will win exactly three times is .294, where

$$f(3) = \binom{5}{3} (18/38)^3 (20/38)^{5-3} = \frac{5!}{3!2!} (18/38)^3 (20/38)^2$$

$$= 10(.10628)(.27701)$$
$$= .294.$$

The probability of success (i.e., red) twice is similarly given by

$$f(2) = \binom{5}{2}(18/38)^2(20/38)^3 = .327.$$

The remaining probabilities f(0), f(1), f(4), f(5) are given by f(0) = .040, f(1) = .182, f(4) = .132 and f(5) = .024. These probabilities are of special interest to Joe because of the strategy in betting that he plans to employ. Joe's plan is to head to the table with $31 worth of chips and to initially bet $1 on red. If successful, the payoff is $1 of profit, which Joe pockets, and he then bets $1 again on red. But, if unsuccessful, Joe loses his dollar bet, and on his next bet he bets $2, double his previous bet. The idea here is that if red comes up, Joe will receive $2 back, which covers his previous dollar loss and furnishes a dollar profit. Again though, if Joe loses his bet, he'll bet $4 on the next spin. This pattern continues with Joe doubling his bet after each loss and betting a dollar after each win. Joe's net earnings for the day increase by one after each win. Unfortunately, the earnings don't increase very rapidly this way, certainly not as rapid as they could decrease. But Joe is a patient fellow, and he has worked out the probabilities for success. Joe's reasoning is that since he has $31 to start with he could suffer a bad string of at most four consecutive losses (this means losing 1+2+4+8 dollars), but with a success on the next spin (with a $16 bet), the net result for the five plays is a dollar profit! Five consecutive losses would cost Joe his $31, thereby leaving him with whatever profit he had accumulated so far and perhaps not enough to bet $32 on the next spin. Now, what are the chances that Joe could lose five consecutive times? Equivalently, what is the probability that red does not come up at all in five spins? This probability is f(0), from the preceding list of probabilities: f(0) = .040. So Joe stands a .04 chance of losing all five spins, or perhaps better sounding, there is a 96 percent chance of the ball landing in a red slot at least once during any five spins of the wheel. Joe therefore feels pretty confident that if he plays for, say, half an hour, he should have accumulated enough profit to at least buy himself a nice dinner.

A Bernoulli process can always be represented graphically by what is known as a tree-diagram. Quite often this helps the student understand the binomial distribution and compute some associated probabilities. The tree-diagram for the roulette wheel will be constructed here. Because of a lack of space on the paper, only three spins of the wheel shall be considered. After the first spin (the first Bernoulli trial) the two outcomes of red (success - S) and non-red (failure - F) occur with probabilities of p(S) = 18/38 and p(F) = 20/38. This is represented in Fig. 6-16. The second spin also yields one of these two possible outcomes, with the same probabilities, but the process of the two spins together yields a total of four (2^2) different outcomes: SS, SF, FS, and FF where, SF represents a success on the first spin and a failure on the second spin. The probability of SF happening is

$$p(SF) = p(S)p(F) = \frac{18}{38} \frac{20}{38} \cdot$$

Figure 6-17 shows the tree-diagram after the third spin.

Notice there are $8 = 2^3$ nodes at the bottom of the diagram. This means there are 8 different paths running down through the diagram, which in turn corresponds to 8 different outcomes for this Bernoulli process. These outcomes can be symbolized by SSS, SSF, SFS, SFF, FSS, FSF, FFS, and FFF. The outcome SSF represents a success on each of the first two spins, and a failure on the third spin: the probability that this outcome happens is $(18/38)^2(20/38)$. Suppose you are interested in having a successful event (i.e., the ball

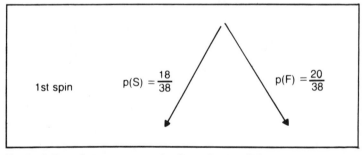

Fig. 6-16. Tree-diagram representing first spin at roulette.

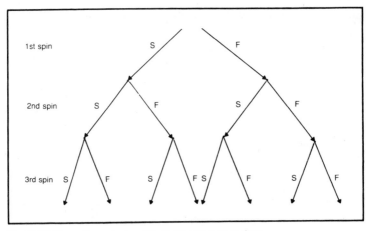

Fig. 6-17. Tree-diagram showing first 3 spins at roulette.

lands in a red slot) exactly twice during the three spins. This means you want two successes and one failure, but it doesn't make any difference what order these occur in. The possibilities are SSF, SFS, and FSS. These three possibilities are represented by the dashed lines in the tree-diagram in Fig. 6-18.

Consequently, the probability of exactly two successes in three spins is given by

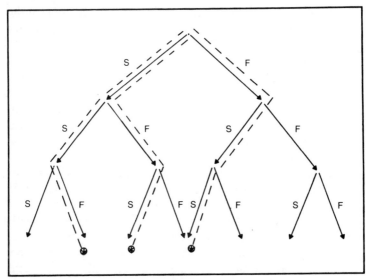

Fig. 6-18. Two successes and one failure.

p(exactly two successes)$=$ p(SSF) + p(SFS) + p(FSS)

$$= \frac{18}{38} \frac{18}{38} \frac{20}{38} + \frac{18}{38} \frac{20}{38} \frac{18}{38} + \frac{20}{38} \frac{18}{38} \frac{18}{38}$$

$$= 3 \left(\frac{18}{38}\right)^2 \frac{20}{38} \cdot$$

This is where the binomial coefficient symbol comes into play because

$$3 \left[\left(\frac{18}{38}\right)^2 \left(\frac{20}{38}\right)\right] = \binom{3}{2}\left(\frac{18}{38}\right)^2 \left(\frac{20}{38}\right)^{3-2} \cdot$$

The binomial coefficient is actually the counter that indicates how many different ways you can obtain the desired number of successes. Thus, if Joe Tourist plays five games of roulette, exactly two successes can occur in any one of $\binom{5}{2} = 10$ different ways:

SSFFF	SFFSF	FSSFF	FSFFS	FFSFS
SFSFF	SFFFS	FSFSF	FFSSF	FFFSS

Two of the more important properties of the general binomial distribution are that the mean of the distribution is $\mu = np$, and the standard deviation is $\sigma = \sqrt{npq}$. So if you suppose that Joe Tourist is a glutton for punishment and embarks on a lengthy half a day of play at the roulette table, for say 800 games, how many times should you expect Joe to win? Since the mean is $np = 800(18/38) = 378.94$, you can expect the ball to land in a red slot approximately 379 times. Of course it may not occur exactly that many times, but the Law of Large Numbers for Bernoulli Trials asserts [76] that this figure is the best approximate.

Closely akin to the binomial distribution is the negative binomial distribution [5], sometimes known as Pascal's distribution. If, as before, you consider a sequence of independent Bernoulli trials, where the probability for success is p, you might be interested in knowing how many trials are needed before the k-th success is reached. Let the random variable X represent this unknown number of trials. Then the negative binomial probability function g gives the probability that $X = i$ by

$$g(i) = \text{Prob}(X = i) = \binom{i-1}{k-1} p^k q^{i-k}$$

where $i = k, k+1, \cdots$. To illustrate, if Joe Tourist wishes to continue play until he accumulates a \$10 profit, the probability that he can do this in 17 games is given by

$$g(17) = \binom{16}{9}\left(\frac{18}{38}\right)^{10}\left(\frac{20}{38}\right)^{7} = .07$$

while the probability that he can do it in 21 games is

$$g(21) = \binom{20}{9}\left(\frac{18}{38}\right)^{10}\left(\frac{20}{38}\right)^{11} = .08.$$

The negative binomial distribution occurs commonly in practical applications, and it can be shown that this form of distribution can result from several sets of assumptions about the underlying situation [5]. Take for example, the distribution of results from accident proneness models in which the accidents occuring in a population of human subjects during a specified period are recorded. If the number of accidents to particular individuals follow a Poisson distribution while the parameter of this distribution, which may be said to measure "accident proneness", varies within the population according to a certain type of continuous distribution, then the total number of accidents in the population follows a negative binomial distribution. The distribution also occurs in studies of the distribution of plants or animals when there is a tendency towards clumping among individual specimens.

The remainder of this section will focus on the binomial probability function. Recall that

$$f(i) = \text{Prob}(X = i) = \binom{n}{i} p^i q^{n-i} = \frac{n!}{i!\,(n-i)!}\, p^i q^{n-i}$$

gives the probability of exactly i successes in a sequence of n independent Bernoulli trials, where p is the probability for an individual success. It is not very difficult to compute $f(i)$ as soon as the values of n, i, and p are known. You simply compute $n!$, $i!$, $(n-i)!$, p^i and $q^{n-i} = (1-p)^{n-i}$ and then form the product

357

$$f(i) = [n!/i!] * [1/(n-i)!] * p^i * q^{n-i}.$$

But, watch out, because the value of n! (or any factorial) grows exceedingly fast. Storage capacity is quickly surpassed as soon as n reaches a value of approximately 12 or 13. When this happens a considerable amount of decimal accuracy is lost. This error is then carried through the rest of the calculations and perhaps is magnified during the process. A more efficient way of calculating f(i) is thus desired. Upon inspection of the binomial coefficient you can see that the cancellation laws from elementary algebra enable you to write

$$\binom{n}{i} = \frac{n!}{i!\,(n-i)!} = \frac{n(n-1)\,(n-2)\,\cdots\,(3)\,(2)\,(1)}{[i(i-1)(i-2)\cdots(2)(1)]\,[(n-i)(n-i-1)\cdots(3)(2)(1)]}.$$

Then you can cancel out either the terms $i, i-1, i-2, \cdots, 2$ or $(n-i)$, $(n-i-1), \cdots, 2$ from both the numerator and the denominator. Choose to cancel out the greater number of terms: this is determined by checking which of i or $(n-i)$ is larger. Thus, you would write

$$\binom{6}{2} = \frac{6 \cdot 5 \cdot 4 \cdot 3 \cdot 2 \cdot 1}{2 \cdot 1 \cdot 4 \cdot 3 \cdot 2 \cdot 1} = \frac{6 \cdot 5}{2 \cdot 1}$$

or

$$\binom{10}{7} = \frac{10 \cdot 9 \cdot 8 \cdot 7 \cdot 6 \cdot 5 \cdot 4 \cdot 3 \cdot 2 \cdot 1}{7 \cdot 6 \cdot 5 \cdot 4 \cdot 3 \cdot 2 \cdot 1 \cdot 3 \cdot 2 \cdot 1} = \frac{10 \cdot 9 \cdot 8}{3 \cdot 2 \cdot 1}.$$

What then is the most effective way of evaluating this binomial coefficient

$$\binom{10}{7}?$$

Should you multiply $10 \cdot 9 \cdot 8$ and divide that by the product $3 \cdot 2 \cdot 1$? Since the numbers involved are small, it doesn't make much difference, but consider the coefficient $\binom{50}{20}$. Your first reduction via cancellation is to express this as

$$\binom{50}{20} = \frac{50 \cdot 49 \cdot 48 \cdots 31}{20 \cdot 19 \cdot 18 \cdots 1}$$

where it is obvious that both numerator and denominator are too large for the computer to handle. This fraction could instead be broken up into a product of 20 other fractions by

$$\frac{50}{20} \frac{49}{19} \frac{48}{18} \cdots \frac{31}{1}$$

which would be much easier to work with. There are several problems that remain. First, this product may still be too large for the computer. Second, computing p^i and q^{n-i} may involve numbers that are too small to be handled effectively. You may wish to give some serious thought now as to how best compute f(i). Perhaps something along the lines shown in the following example might prove advantageous.

$$\binom{50}{20} p^{20} q^{30} = \frac{50 \cdot 49 \cdot 48 \cdots 31}{20 \cdot 19 \cdot 18 \cdots 1} \; p^{20} q^{30}$$

$$= \left[\frac{50}{20} \, pq \right] \left[\frac{49}{19} \, pq \right] \cdots \left[\frac{31}{1} \, pq \right] q^{10} \, .$$

In passing, it might be helpful for you to know that in many instances when n is large, the binomial probability is estimated by another probability distribution, namely the normal distribution [41]. This is a powerful technique since it allows for a transition from a discrete to a continuous random variable.

From an amusing and recreational sidelight, you may wish to consult [47] for an enjoyable application of the binomial distribution to the great American game of baseball.

Exercises

1. Write a program that computes the value of the binomial probability f(i) when the values of n, i and p are given. Let us assume n is not larger than 50.
2. Stirling's formula for an approximation to n! is given by

$$n! \doteq n^n e^{-n} \sqrt{2\pi n}.$$

If k is an arbitrary positive integer, use Stirling's formula to obtain an approximate to the binomial coefficient

$$\binom{kn}{n}.$$

CENTRAL LIMIT THEOREM

A substantial portion of statistics lies in estimation. The student is customarily exposed to certain properties of estimators and to general methods of estimation. Perhaps the most commonly asked question and surely one of the most important is "What confidence do I have that the estimate I make will be close to the true value of the parameter in question?" Alternately put, you are concerned that the difference between the estimate $\hat{\theta}$ and the parameter θ, $|\hat{\theta} - \theta|$ is smaller than some prescribed tolerance. Now, you can answer these questions by knowing the probability distribution of the estimator for each value of the parameter. Unfortunately, this is not always known. On the other hand, for the special situation of estimating the population mean, the Central Limit Theorem proves indispensble. First proved by Lindeberg and Levy in 1925, this theorem is one of the most important results of probability theorem [41].

Central Limit Theorem: If all possible random samples, each of siz‿ n, are taken from any population with a mean μ and standard deviation σ, the sampling distribution of sample means $\bar{x}$ will have a mean $\mu_{\bar{x}}$ equal to μ, a standard deviation $\sigma_{\bar{x}}$ equal to $\sigma/\sqrt{n}$, and will be approximately normally distributed.

In essence, if the parent population is normally distributed to begin with, the distribution of the sample means will be normally distributed; but if the parent population is not normal, the sampling distribution will be approximately normally distributed for large enough sample sizes—usually 30 or more. The approximation to the normal distribution improves as the sample size n increases.

The Central Limit Theorem is valuable because it supplies three important pieces of information about the sampling distribution of sample means of size n, namely

360

1. the center of the distribution.
2. the magnitude of the dispersion
3. how it is distributed

It might be of value to consider a particular example, and show just what the Central Limit Theorem has to say. Consider as the parent population the four element set $\{1,2,3,4\}$, where each of these elements occurs with equal probability (namely $p = .25$). This is not a normal distribution, but rather a constant distribution whose graph is depicted in Fig. 6-19. The population mean μ and standard deviation σ can be found by using the formulas

$$\mu = \Sigma \; xP(x)$$
$$\sigma = [\; \Sigma \; x^2P(x) - (\; \Sigma \; xP(x))^2]^{1/2}.$$

Thus, as the results from Table 6-10 show, $\mu = 2.5$ and $\sigma = \sqrt{5}/2$.

Now consider all possible samples of size 2 (n = 2) from $\{1,2,3,4\}$. These are all the possible samples when the elements are drawn, replaced, and possibly redrawn. These 16 samples and their sample means $\overline{x}$ are as shown in Table 6-11. The mean $\mu_{\overline{x}}$ and standard deviation $\sigma_{\overline{x}}$ for the distribution of sample means $\overline{x}$ is computed as $\mu_{\overline{x}} = 2.5$ and $\sigma_{\overline{x}} = \sqrt{10}/4$ as shown in Table 6-12. According to the Central Limit Theorem, these values must satisfy the relationship $\mu_{\overline{x}} = \mu$ and $\sigma_{\overline{x}} = \sigma/\sqrt{n}$, which they do because

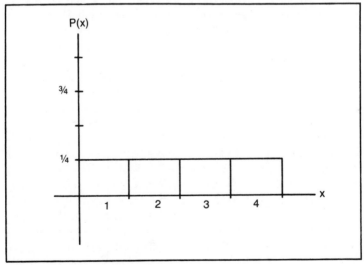

Fig. 6-19. A constant distribution.

Table 6-10. Computing a Mean and a Standard Deviation.

x	P(x)	xP(x)	$x^2 P(x)$
1	1/4	1/4	1/4
2	1/4	2/4	4/4
3	1/4	3/4	9/4
4	1/4	4/4	16/4
		10/4	30/4

$$\mu = 10/4 = 2.5$$

$$\sigma = \sqrt{30/4 - (10/4)^2} = \sqrt{5}/2$$

$$\sigma_{\bar{x}} = \sqrt{10}/4 = \sqrt{5}\sqrt{2}/4 = \sqrt{5}/(2\sqrt{2}) = \sigma/\sqrt{n}.$$

The Central Limit Theorem further specifies the nature of the distribution of the sample means $\bar{x}$, namely that it is approximately normal. The distribution, pictured in Fig. 6-20, is a symmetric distribution that has the appearance of being close to a normal distribution. The resemblance would be much greater if the sample size were larger.

For practical applications, the real life situation does not involve a sampling distribution of all the possible samples of size n, but rather some proper subset. The immediate implication here is that the mean of the sample means will probably not be equal to the population mean μ, and the sample standard deviation will probably

Table 6-11. Samples of Size 2 from {1,2,3,4}.

Samples	$\bar{x}$	Samples	$\bar{x}$
1,1	1	3,1	2
1,2	3/2	3,2	5/2
1,3	2	3,3	3
1,4	5/2	3,4	7/2
2,1	3/2	4,1	5/2
2,2	2	4,2	3
2,3	5/2	4,3	7/2
2,4	3	4,4	4

Table 6-12. Computing Sample Mean and Standard Deviation.

$\overline{x}$	$P(\overline{x})$	$\overline{x}P(\overline{x})$	$\overline{x}^2 P(\overline{x})$
1	1/16	1/16	1/16
3/2	2/16	3/16	9/32
2	3/16	6/16	12/16
5/2	4/16	10/16	25/16
3	3/16	9/16	27/16
7/2	2/16	7/16	49/32
4	1/16	4/16	16/16
		40/16	110/16

$$\mu_{\overline{x}} = 40/16 = 2.5$$

$$\sigma_{\overline{x}} = \sqrt{110/16 - (40/16)^2} = \sqrt{10}/4.$$

be different from $\sigma/\sqrt{n}$. But, the Central Limit Theorem can also be worded so as to guarantee that these values are approximately equal; in other words, you have

$$\mu_{\overline{x}} \doteq \mu$$
$$\sigma_{\overline{x}} \doteq \sigma/\sqrt{n}$$

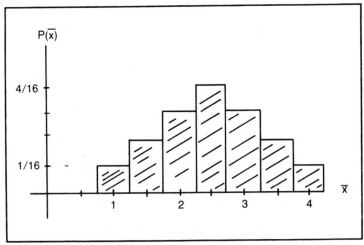

Fig. 6-20. A distribution of sample means.

for any random sample drawn, provided the sample size is large. When the parent population sample has a normal probability distribution, the sampling distribution of $\bar{x}$ is exactly normal for any random sample.

Once again, this approximation can be demonstrated by considering the elements $\{1,2,3,4\}$ distributed evenly. Thirty-six samples of size 3 are drawn, and these are listed in Table 6-13, together with their means. The observed mean of these sample means is computed to be $263/108 = 2.435$, which is within .065 of the population mean $\mu = 2.5$. The sample standard deviation $s_{\bar{x}}$ is

$$s_{\bar{x}} = \sqrt{\frac{2069}{324} - \left(\frac{263}{108}\right)^2}$$
$$= .675$$

Table 6-13. Sample of Size 3.

Sample	$\bar{x}$	Sample	$\bar{x}$
1,1,1	3/3	2,1,1	4/3
1,1,2	4/3	2,1,3	6/3
1,2,1	4/3	2,1,4	7/3
1,2,2	5/3	2,2,1	5/3
1,2,3	6/3	2,2,3	7/3
1,3,1	5/3	2,3,1	6/3
1,3,4	8/3	2,3,2	7/3
1,4,2	7/3	2,4,1	7/3
1,4,4	9/3	2,4,4	10/3

Sample	$\bar{x}$	Sample	$\bar{x}$
3,1,2	6/3	4,1,1	6/3
3,1,3	7/3	4,1,4	9/3
3,1,4	8/3	4,2,1	7/3
3,2,2	7/3	4,2,2	8/3
3,2,3	8/3	4,3,2	9/3
3,2,4	9/3	4,2,3	9/3
3,4,2	9/3	4,3,3	10/3
3,4,3	10/3	4,4,1	9/3
3,4,4	11/3	4,4,3	11/3

which is within .030 of the value $\sigma/\sqrt{3}$.

A most important application of the Central Limit Theorem concerns making probability statements about the sample means. You know that the sampling distribution is approximately normal, with a mean of μ and a standard deviation of $\sigma/\sqrt{n}$. This distribution is then standardized by subtracting μ from all the values of $\bar{x}$, and dividing by $\sigma/\sqrt{n}$. This new variable is denoted by z, so

$$z = \frac{\bar{x} - \mu}{\sigma/\sqrt{n}}.$$

The distribution of z scores is then normally distributed, but with a mean of 0 and a standard deviation of 1. This follows from properties of the expectation and variance operator [56]. Questions concerning the distribution of $\bar{x}$ scores can thus be reworded to involve the z scores. Consider the following example.

> Example: It is known that the heights of sixth grade children are normally distributed with a mean $\mu = 54$ inches and standard deviation $\sigma = 3$. Mrs. Jones' sixth grade class has 36 pupils. What is the probability (likelihood) that the class average height $\bar{x}$ is somewhere between 52.5 and 55 inches?

What you wish to find is the probability, $P(52.5 < \bar{x} < 55)$. Since $\bar{x} = 55$ corresponds to a z score of $z = (55 - 54)/(3/\sqrt{36})$, and $\bar{x} = 52.5$ corresponds to $z = (52.5 - 54)/.5$, then

$$P(52.5 < \bar{x} < 55) = P\left(\frac{52.5 - 54}{3/\sqrt{36}} < z < \frac{55 - 54}{3/\sqrt{36}}\right)$$

$$= P(-3 < z < 2).$$

This probability involving a z score can be found in a standard table in the back of most statistics books.

For another example, suppose that the amount of time in minutes devoted to commercials on a TV channel during any half hour program is a random variable X whose mean is 6.3 minutes and whose standard deviation is 0.866 minutes. What is the approximate probability that a person who watches 40 half hour programs will be exposed to over 250 minutes of commercials?

For the solution to this, note that $\mu = 6.3$, $\sigma = 0.866$, and $n =$

40. Total commercial time of 250 minutes implies an average of 250/40=6.25 minutes per half hour program. So you are interested in $P(\bar{x} > 6.25)$. The Central Limit Theorem gives us $\mu_{\bar{x}} = 6.3$ and $\sigma_{\bar{x}} = 0.866/\sqrt{40} = .1369$; thus

$$P(\bar{x} > 6.25) = P\left(z > \frac{6.25 - 6.3}{.1369}\right)$$

$$= P(z > -.365)$$

$$= .6424$$

where this last figure is taken from a table of z-values.

All of the discussion and examples presented so far have been directed at discrete probability distributions. The Central Limit Theorem assures you that as the sample size increases, the distribution of sample means approaches a normal distribution. A normal distribution is a continuous distribution, not a discrete distribution, that is described by a curve whose mathematical equation is

$$y = \frac{1}{\sqrt{2\pi}} e^{-x^2/2}$$

and that is graphed as shown in Fig. 6-21. Consequently, if $\bar{x}$ is the sample mean of a size n from a parent population with mean μ and

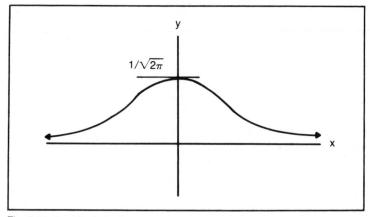

Fig. 6-21. A normal distribution curve.

standard deviation σ, the Central Limit Theorem can be reformulated (using concepts from calculus) to state that for any two real numbers a, b with a > b, you have

$$\lim_{n \to \infty} P\left(a < \frac{\bar{x} - \mu}{\sigma/\sqrt{n}} < b\right) = \frac{1}{\sqrt{2\pi}} \int_a^b e^{-x^2/2} \, dx.$$

You are referred to [12] for further discussion of this nature.

Exercises

1. Write a program that will help to verify part of the conclusion to the Central Limit Theorem. Feed into the program an integral value of k with k ⩾ 10, and then consider the elements from the set $X = \{1,2,3,\cdots,k\}$. This will imply (won't it?) that $\mu = k(k+1)/2$. What is the value of σ? Then, for a given value of n, with $1 < n < k$, write a program that generates all samples of size n from X. Compute $\bar{x}$ for each sample. Compile this data and form a frequency table. Compute the mean $\mu_{\bar{x}}$ and the standard deviation $\sigma_{\bar{x}}$, and show that these values obey the relationship

$$\mu_{\bar{x}} = \mu$$
$$\sigma_{\bar{x}} = \sigma/\sqrt{n}$$

as described in the Central Limit Theorem. Test your program with the previously mentioned example of $X = \{1,2,3,4\}$ and n = 2.

2. How many samples of size n < k are there from a parent set of $X = \{1,2,3,\cdots,k\}$ if it is stipulated that each sample must contain n different elements?

THE BIRTHDAY PROBLEM

A course on probability and statistics presents an instructor with a multitude of topics that can be cleverly presented to the student in a fashion so as to arouse considerable interest. Man's fascination with gambling, betting, and using household statistics, all contribute to the ease of this arousal. Occasionally, though, a seemingly unimportant problem which is attractive enough to withstand the typical student's rebellious attitude toward the subject may arise. I have found the birthday problem always fits into this scheme. It has always been a joy to present this to a class. Considerable excitement always is present and an important lesson in probability can be presented. The problem is stated as follows.

The Birthday Problem: With n people in a group, what is the probability that at least two of them have the same birthday?
(This means same day and month, but not necessarily the same year.)

A typical class of size 25 to 30 is a perfect setting for the instructor, because the students are apt to think the chances are slim two of their members have the same birthday; meanwhile the instructor know the exact odds, which in effect are quite good as you shall soon see.

From a programming standpoint, it would not be difficult for you to gather empirical data on the birthday probability. For each given value of n, the class size, you would generate n integral random numbers i of magnitude $1 \le i \le 365$. Each value of i represents one of the days of the year. Thus, $i = 1$ represents January 1, $i = 2$ represents January 2, $i = 32$ represents February 1, and $i = 365$ represents December 31. Either all n of these values are different, or two of them are the same, which means there is a birthday matchup. The ratio of the number of matchups to the number of trial runs would then give you an estimate of the birthday probability. The following program reflects these points.

Program

```
100   REM PROGRAM ON THE BIRTHDAY PROBLEM
110   DIM BIRTHDAY(100)
120   INPUT TRIALRUNS, N
130   REM THE CLASS SIZE IS N
140   MATCHUPS = 0
150   FOR COUNT = 1 TO TRIALRUNS
160       FOR I = 1 TO N
170           BIRTHDAY(I) = INT(365*RND + 1)
180       NEXT I
190       REM NOW TO SEE IF THERE ARE ANY
          MATCHUPS
200       FOR J = 2 TO N
210         FOR I = 1 TO (J−1)
220           IF BIRTHDAY(J) = BIRTHDAY(I) GOTO 260
```

```
230              MATCHUPS = MATCHUPS + 1
240              GO TO 280
250            NEXT I
260          NEXT J
270          REM NO MATCHUPS
280      NEXT COUNT
290      RATIO = MATCHUPS/TRIALRUNS
300      PRINT "OUR ESTIMATE FOR THE BIRTHDAY
         PROBABILITY IS"; RATIO
310      STOP
320      END
```

It is not too difficult, and in fact, it is quite instructive to discuss the derivation of the birthday probability formula. To this end, let E denote the event that at least two people from a class of size n have the same birthday. The complement E^c is the event that everyone has a different birthday. The probabilities of these events sum to one,

$$p(E) + p(E^c) = 1.$$

As soon as you compute the probability of E^c, you will be able to determine $p(E)$. First let E_i be the event that the i-th person (the students have been sequenced 1,2,•••,n) does not have a birthday in common with the preceding (i-1) people; i.e., the i-th person does not have a birthday in common with any of his predecessors. Consequently, the probability of E_1 is 1 ($p(E_1) = 1$), since the first person has no predecessors. The probability of E_2 is 364/365 since there are 364 possible days for which the second person could have a birthday different from the first person. Continuing in this manner yields the probabilities,

$$P(E_3) = 363/365$$
$$p(E_4) = 362/365$$
$$\cdot$$
$$\cdot$$
$$\cdot$$
$$p(E_{10}) = 356/365$$

with the general expression being,

369

$$p(E_i) = \frac{365 - (i-1)}{365}.$$

Note that the event E^c is the intersection of the events $E_1, E_2, \cdots, E_n$,

$$E^c = E_1 \cap E_2 \cap E_3 \cap \cdots \cap E_n$$

and because the events E_i are all independent, this gives the probability of E^c as

$$p(E^c) = p(E_1)p(E_2)p(E_3)\cdots p(E_n)$$

$$= \frac{365}{365} \quad \frac{364}{365} \quad \frac{363}{365} \quad \cdots \quad \frac{365 - (n-1)}{365}$$

and thus the desired birthday probability $p(E)$ is

$$p(E) = 1 - p(E^c)$$

$$= 1 - \left[\frac{364}{365} \quad \frac{363}{365} \quad \cdots \quad \frac{365 - (n-1)}{365}\right].$$

Values of $p(E)$ for various n are listed in the Table 6-14. Surprisingly enough, as soon as n exceeds 22, the probability $p(E)$ exceeds .50, which means that for a class of size 23 or larger, the chances are better than 1 in 2 that there will be two people in the class with the same birthday!

Some very interesting generalizations of the birthday problem have been studied [1,52]; two of these are listed on the following page.

n	p(E)
5	.027
10	.117
15	.253
20	.411
25	.569
30	.706
35	.814
40	.891

Table 6-14. Values of the Birthday Probability.

Problem 1. Of n people in a group, what is the probability that at least k (k ≥ 2) of them have the same birthday?

Problem 2. Suppose a year has D days. Of n people in a group, what is the probability that the birthdays of every pair are at least k (k ≥ 1) days apart?

You know from above that a minimal group size of 23 is needed before the probability exceeds .50 that at least two people have the same birthday. The solution to Problem 1 shows us that a minimal group size of 88 is needed for at least three (k = 3) people to have common birthdays, and a group of 187 for k = 4. These figures make use of the probability formula

$$P = \frac{n!}{\prod\limits_{i=1}^{k-1}(n_i!)(i!)^{n_i}} \; \frac{Perm(365, \sum\limits_{i=1}^{k-1} n_i)}{365^n}$$

where n_i represents the number of (i−1)-tuple birthdays.

n	P
8	.208
9	.260
10	.314
11	.370
12	.426
13	.482
14	.537
⋮	
24	.908

Table 6-15. Values of the Generalized Birthday Probability.

The second problem is intriguing not only because it varies the length of the year, but also because it is concerned with birthdays that are close together. For instance, if D = 365 and k = 1, you have the same situation as the ordinary birthday problem; but if k = 2, the problem can be restated so that it concerns the probability P that some pair have birthdays that are either the same or differ by one day. Sample results in this case include the data listed in Table 6-15. It is apparent that for a class of at least 14 the odds of finding a pair of students with birthdays that differ by at most one day are favorable, which means the probability is at least .5. The odds are very

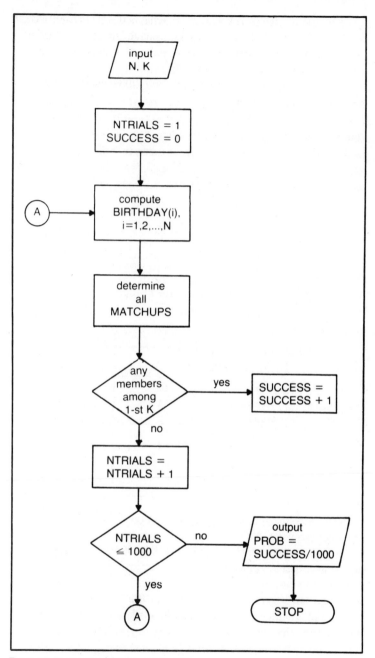

Fig. 6-22. Estimating birthday probability.

favorable (P = .908) in a class of 24. Furthermore, if D = 365 and k = 3, the probability that a pair of students have birthdays that differ by at most two days is at least .5 as soon as n attains the value 11. That's certainly not a very large class.

If you reduce the year to include only 200 days, D = 200, then for a class of 11, the probability is approximately .25 that some pair have the same birthday (k = 1), while a class of at least 16 is required before the probability reaches .50. These figures are found by evaluating the formula

$$P = 1 - \frac{(n-1)!}{D^{n-1}} \left(\begin{array}{c} D - n(k-1) - 1 \\ n-1 \end{array} \right)$$

where the large parentheses denotes the binomial coefficient.

All of these generalized birthday problems lend themselves very well to computer simulation for estimating probabilities. Basically, all that is involved is to generate some random integers, to search through these for a pair or triple that are either equal or nearly equal, and to keep track of the number of successes. This method will be applied to the following birthday problem [32,58].

Problem 3. Let k ⩾ 1 be given. Of a group of n people, what is the probability P(n,k) that at least two people have the same birthday, with at least one of the two being among the first k people?

The flowchart shown in Fig. 6-22 outlines how to estimate this probability by running a series of 1000 trials.

Exercises

1. Write a program that approximates the value of P(n,k). Assume that n ⩽ 100 and k ⩽ 25. Try and find the value of n for which p(n,10) ≐ .5.
2. If a year has 100 days, what is the size of the smallest class needed to insure a probability of at least 1/2 that a pair of students can be found whose birthdays are at most 1 day apart?

Appendix A
Answers to Problems

The following are the answers to problems that form the second part
of each set of exercises.

Name of Section   **Title**

Simplifying   Just $b = 4$.
Fractions

Negative   (a) 23_{-10} (b) 0_{-10} (c) 197_{-10}
Bases

Bachet's   Label the coins 1 through 12. On the first
Weights   weighing, coins 1-4 are weighed against coins
                   5-8. Either the scale balances, or it doesn't. If it
                   balances, then weigh coins 9,10,11 against
                   2,3,4. If it doesn't balance, then weigh coins
                   1,2,5 against 3,6,7. If this doesn't balance,
                   weigh 1,6 against 2,7; if it does balance, weigh
                   1 against 4.

Lattice $A = 19 + 24/2 - 1 = 30$.
Points

Primes $10^{10^{34}} + 1$

Since $\log_{10}(M_{44497}) = 13394.9$, it follows that prime contains 13395 digits.

Wierd Numbers	945
Euclid's Algorithm	$m = 54$, $w = 28$.
Powerful Numbers	1/2
Fermat's Theorem	The last 3 digits are 384 because $2^{1314} \equiv 384$ (mod 1000).
Quadratic Congruences	There are no numbers of the form $n^2 + 1$ that are divisible by 7, because $n^2 + 1 \equiv 0$ (mod 7) is equivalent to $n^2 \equiv -1$ (mod 7), and $(-1)^{(7-1)/2} \equiv -1$ (mod 7).
Quadratic Reciprocity	If $p \equiv 1$ (mod 8) then $p = 8k + 1$. Thus $(p^2 - 1)/8$ equals $8k^2 + 2k$, which is an even number; so $(-1)^{(p^2-1)/8} = 1$. Likewise for $p \equiv 7$ (mod 8), $p \equiv 3$ (mod 8) and $p \equiv 5$ (mod 8).
Descartes's Rule of Signs	Yes, because this is same equation as $x^4 - x^2 + 4x = 0$, or $x(x^3 - x + 4) = 0$. For $x^3 - x + 4$, $V^+ = 2$, $V^- = 1$, $M = 0$, but $x^3 - x + 4$ has no positive roots. Hence there are 2 complex roots.
Luddhar's Method	$l = 2$, $m = 1$, $p = 2$, $q = 1$.
Quartic Equations	$(.4142, .1716)$ and $(-2.4142, 5.8284)$
Budan and Sturm Methods	Using Budan's method, $V_5 = 0$ and $V_r = 0$ for $r > 5$.

Conics	The equation of the ellipse is $x^2/100 + y^2/36 = 1$. At the point $P(6,q)$ we have $6^2/100 + q^2/36 = 1$, so $q = 4.8$. Thus, the height is $12 + 4.8 = 16.8$.
Countability of Rational Polynomials	Since $9{,}507{,}960 = 2^3 3^2 5^1 7^4 11^1$ this gives SUM $= 3$, LCD $= 2$, $b_0 = 1$, $b_1 = 4$, $b_2 = 1$ and $n = 2$. Then $s_2 = 0$, $s_1 = 1$ and $s_0 = 1$; so $a_0 = -1/2$, $a_1 = -2$ and $a_2 = 1/2$. The polynomial is $p(x) = (1/2)x^2 - 2x - (1/2)$.
Pivoting	$x_1 = -\beta/(1 - \beta)$ and $x_2 = (2 - \beta)/(1 - \beta)$. When $\beta = 1 - 10^{-10}$ the solutions are approximately $x_1 = 1 - 10^{10}$ and $x_2 = 1 + 10^{10}$.
Splines	$f(x) = (1/2)(x^4 - 2x^3 - 3x^2 + 2x + 4)$.
Matrix Square Roots	$b = (\lambda_1 + \lambda_2 - 1 - \lambda_1\lambda_2)/2$ and $d = \lambda_1 + \lambda_2 - 1$.
The Gram-Schmidt Process	$S' = \left\{1, \sqrt{12}\,(x - 1/2), \sqrt{180}\,(x^2 - x + 1/6)\right\}$.
Sequences	$e^{\sqrt{2}\,\pi}$
Differentiation	$f(x) = (x - 1)^{10/3}$
Directional Derivatives	

$$D_u f(a,b,c) = \frac{\delta f}{\delta x}\,(a,b,c) \bullet u_1 + \frac{\delta f}{\delta y}\,(a,b,c) \bullet u_2 + \frac{\delta f}{\delta z}\,(a,b,c) \bullet u_3$$

where

$$u = u_1 i + u_2 j + u_3 k \text{ and } u_1 = \frac{d - a}{\sqrt{(d-a)^2 + (e-b)^2 + (g-c)^2}},$$

$$u_2 = \frac{e - b}{\sqrt{(d-a)^2 + (e-b)^2 + (g-c)^2}}$$

377

$$, u_3 = \frac{g-c}{\sqrt{(d-a)^2 + (e-b)^2 + (g-c)^2}}$$

Fixed Points	$\sqrt{3/2},\ \sqrt{(k/(k-1))}$

Locating Roots	$2,\ 2 \pm \sqrt{(5 \pm \sqrt{5})/2}$

Numerical Integration

$$A_r = (2/N)\,[\,f(1/N) + f(3/N) + \cdots + f((2N-1)/N)]$$
$$= (2/N)\,[\,3N + 2N(2N+1)(2N-1)/6N^2\,]$$
$$= (26N^2 - 2)/(3N^2).$$

Romberg Integration

$$[64A_s(2n,4n) - A_s(n,2n)]/63$$

The Beta Function

$$\frac{\partial}{\partial x} B(x,y) = \frac{\partial}{\partial x}\left[\int_0^1 t^{x-1}(1-t)^{y-1}dt\right] =$$

$$\int_0^1 (1-t)^{y-1}\,\frac{d}{dx}\,t^{x-1}\,dt = \int_0^1 (1-t)^{y-1}\ln(t)t^{x-1}\,dt$$

$$\frac{\partial}{\partial y} B(x,y) = \frac{\partial}{\partial y}\left[\int_0^1 t^{x-1}(1-t)^{y-1}dt\right] =$$

$$\int_0^1 t^{x-1}\,\frac{d}{dy}\,(1-t)^{y-1}\,dt = \int_0^1 t^{x-1}\ \ln(1-t)(1-t)^{y-1}dt$$

The Zeta Function

sum $= \pi/4$ (let $\alpha = \pi/2$).

La Grange Multipliers

$$\begin{cases} yz = 1 \cdot \lambda \\ xz = 3 \cdot \lambda \\ xy = 4 \cdot \lambda \end{cases} \rightarrow \lambda = 108 \text{ at } (36,12,9).$$

Euler's Constant

$f'(x) = -x^{-2}(x+1)^{-1} < 0$ which says f is decreasing. Furthermore, $f(x) \rightarrow 0$ as $x \rightarrow \infty$. Since $f(1) = 1 - \ln(2) > 0$, then $f(x) > 0$ for $x \geq 1$.

Euler's Method	$y_n = (1 + 1/n)^n$ and $\lim y_n = e$.
Expected Value	$36/14$.
Measures of Central Tendency	$1, 2, 3, 9, 9$
Means	Square both sides; simplify; and square again. This yields $(x_1 y_2 - x_2 y_1)^2 \geq 0$.
Buffon's Needle Problem	One
The Line of Best Fit	The line LOMBF is a horizontal line $(m = 0)$ or a vertical line.
The Quadratic of Best Fit	$A = 1, B = 2, C = 3$: $u = 0, v = 0$, so you won't be able to compute A, computer will display an error message.
Chebyshev's Inequality	$\mu = 3, \sigma = \sqrt{2}, k = \sqrt{2}, 1/k^2 = 1/2$. Thus $P(|X - 3 \geq 2) \leq 1/2$.
Binomial Distribution	$\dbinom{kn}{n} \doteq \left[\dfrac{k^k}{(k-1)^{k-1}}\right]^n * \sqrt{\dfrac{k}{k-1}} * \dfrac{1}{\sqrt{2\pi n}}$
General Limit Theorem	$\dbinom{k}{n} = \dfrac{k!}{n!\,(k-n)!}$
The Birthday Problem	$n = 8$, for if $D = 100, k = 2$ and $n = 8$ then $P = 1 - \dfrac{7!}{100^7}\dbinom{91}{7} = .6$. If $n = 7$, then $P = 1 - \dfrac{6!}{100^6}\dbinom{92}{6} = .486$.

Appendix B
References

1. Abramson, Morton and W. Moser. "More birthday surprises," *American Mathematics Monthly*, 77 (1970), 856-858.
2. Apostol, Thomas, "Another elementary proof of Euler's formula for $\gamma(2n)$," *American Mathematics Monthly*, 80 (1973), 425-431.
3. Artin, Emil. *The Gamma Function*. New York, Holt, Rinehart and Winston, 1964.
4. Ashford, John. *Statistics for Management* (2nd ed.,) Renouf, Brookfield, 1981.
5. Ayoub, Raymond. "Euler and the Zeta function," *American Mathematics Monthly*, 81 (1974), 1067-1086.
6. Bateman, Harry. "Halley's method for solving equations," *American Mathematics Monthly*, 45 (1938), 11-17.
7. Bernot, Bruce. "Elementary evaluation of $\gamma(2n)$," *Mathematics Magazine*, 48 (1975), 148-154.
8. Boas, R.P. "Anomalous cancellation," *Mathematical Plums* (The Dolciani Mathematical Expositions - No. 4), *Mathematics Association of America*, (1979), 113-129.
9. Boas, Ralph. *A Primer of Real Functions*. Mathematics Associations of America, Washington, 1960.
10. Book, S. and L. Sher. "How close are the mean and the median?" *Two-Year College Math. J.*, 10 (1979), 202-204.
11. Brown, Ezra. "The first proof of the quadratic reciprocity law, revisited," *American Mathematics Monthly*, 88 (1981), 257-264.
12. Brunk, H.D. *An Introduction to Mathematical Statistics* (2nd ed.,) Blaisdell, New York, 1965.
13. Burden, Richard; Faires, J. and A. Reynolds. *Numerical Analysis* (2nd ed.,) Prindle-Weber-Schmidt, Boston, 1981.
14. Bylinsky, Gene. "The Japanese chip challenge," *Fortune*. March 23, 1981. 115-122.
15. Cheney, Ward and D. Kincaid. *Numerical Mathematics and Computing*. Brooks/Cole, Belmont, 1980.

16. Chong, Kong-Ming. "The arithmetic mean-geometric mean inequality; a new proof," *Mathematics Magazine*, 49 (1976), 87-88.

17. Conkwright, Nelson. *Introduction to the Theory of Equations*. Ginn and Co., Boston, 1941.

18. Dence, Joseph. *Mathematical Techniques in Chemistry*. Wiley, New York. 1975.

19. Dence, Thomas. "Differentiable points of the generalized Cantor function," *Rocky Mt. J.*, 9 (1979), 239-249.

20. — — —. "On the monotonicity of a class of exponential sequences," *American Mathematics Monthly*, 88 (1981), 341-344.

21. ———. *The Fortran Cookbook*. TAB BOOKS, Blue Ridge Summit, 1980.

22. Drucker, Daniel. "A second look at Descartes' rule of signs, " *Mathematics Magazine*, 52 (1979), 237-238.

23. Duncan, R. and M. Weston-Smith. *The Encyclopedia of Ignorance*. Pergamon Press Ltd., Oxford, 1977. 193-194.

24. Ebey, S. and J. Beauchamp. "Larval fish, power plants, and Buffon's needle problem," *American Mathematics Monthly*, 84 (1977), 534-541.

25. Falk Ruma. "Another look at the mean, median, and standard deviation," *Two-Year College Math. J.*, 12 (1981), 207-208.

26. Frank, Alan. "The countability of the rational polynomials: a direct method," *American Mathematics Monthly*, 87 (1980), 810-811.

27. Fulks, Watson. *Advanced Calculus* (2nd ed.,) Wiley, New York, 1969.

28. Gagola, Gloria. "Progress on primes," *Mathematics Magazine*, 54 (1981), 43.

29. Gardner, Martin. "Mathematical Games," *Scientific American*, Sept. (1976), 209-210.

30. ———. "Mathematical Games," *Scientific American,* Aug., (1977), 120-124.

31. Gilbert, W. and R. Green. "Negative based number systems," *Mathematics Magazine*, 52 (1979), 240-244.

32. Goldberg, Samuel. "A direct attack on a birthday problem," *Mathematics Magazine*, 49 (1976), 130-131.

33. Golomb, Solomon. "Powerful numbers," *American Mathematics Monthly*, 77 (1970), 848-852.

34. ———. "The evidence for Fortune's conjecture," *Mathematics Magazine*, 54 (1981), 209-210.

35. Halmos, Paul. "The legend of John von Neumann, *American Mathematics Monthly,* 80 (1973), 382-394.

36. Hardy, G.H. and E.M. Wright. *An Introduction to the Theory of Numbers* (5th ed.) Oxford University Press, Oxford, 1979.

37. Honsberger, Ross. *Mathematical Gems I*. Mathematics Associations of America, Washington, 1973.

38. ———. *Mathematical Gems II*. Mathematics Association of America, Washington, 1976.

39. ———. "Short stories in number theory," *Two-Year College Math. J.* 12 (1981), 36.

40. Horton, Goldie. "A note on the calculation of Euler's constant," *American Mathematics Monthly*, 23 (1916), 73.

41. Johnson, Robert. *Elementary Statistics* (3rd ed.) Duxbury Press, North Scituate, 1980.

42. Kasner, E. and James Newman. *Mathematics and the Imagination*. Simon and Schuster, New York, 1962.

43. Klamkin, M.S. "On the uniqueness of the distribution function for the Buffon needle problem," *American Mathematics Monthly* 60 (1953), 677-680.

44. Kolata, Gina Bari. "Testing for primes gets easier," *Science 209* (1980), Sept. 26, 1503-1504.

45. Ledermann, Walter. *Handbook of Applicable Mathematics III: Numerical Methods.* Wiley, Chichester, 1981.

46. Lehmer, D.H. "On the converse of Fermat's theorem," *American Mathematics Monthly,* 43 (1936), 347-354.

47. Levin, Eugene. "Binomial baseball," *Two-Year College Math. J.,* 12 (1981), 260-265.

48. Levinger, Bernard. "The square root of a 2 by 2 matrix," *Mathematics Magazine,* 53 (1980), 222-224.

49. Liu, Andy, "Lattice points and Pick's Theorem," *Mathematics Magazine,* 52 (1979), 232-235.

50. Luthar, R.S. "Luddhar's method of solving a cubic equation with a rational root," *Two-Year College Math. J.,* 11 (1980), 107-110.

51. Manvel, Bennet. "Counterfeit coin problems," *Mathematics Magazine,* 50 (1977), 90-91.

52. McKinney, E.H. "Generalized birthday problem," *American Mathematics Monthly,* 73 (1966), 385-387.

53. Melzak, Z.A. *Companion to Concrete Mathematics.* Wiley, New York, 1973.

54. Moore, Hal and Adil Yaqub. *Elementary Linear Algebra with Applications.* Addison-Wesley, Reading, 1980.

55. Moser, Leo. "On the series Σ 1/p," *American Mathematics Monthly* 65 (1958), 104-105.

56. Neter, J., Wasserman, W. and G. Whitmore. *Applied Statistics.* Allyn and Bacon, Boston, 1978.

57. Newman, James. *The World of Mathematics.* Vol 1. Simon and Schuster, New York, 1956.

58. Nymann, J. "Another generalization of the birthday problem," *Mathematics Magazine,* 48 (1975), 46-47.

59. Ogilvy, C.S. "The beta function," *American Mathematics Monthly,* 58 (1951), 475-479.

60. Ore, Oystein. *Number Theory and its History.* McGraw Hill, New York, 1948.

61. Papadimitriou, Ioannis. "A simple proof of the formula $\Sigma\, k^{-2} = \pi^2/6$," *American Mathematics Monthly,* 80 (1973), 424-425.

62. Polya, G. and G. Szego. *Problems and Theorems in Analysis* I. Springer-Verlag, New York, 1972.

63. Prenter, P.M. *Splines and Variational Methods.* Wiley, New York, 1975.

64. "Problems Section." No. 136. *Two-Year College Math. J.,* 11 (1980), 276.

65. Ramaley, J. "Buffon's needle problem," *American Mathematics Monthly,* 76 (1969), 916-918.

66. Rao, S.K. Lakshmana. "On the sequence for Euler's constant, *American Mathematics Monthly,* 63 (1956), 572-573.

67. Rudin, W. *Principles of Mathematical Analysis.* McGraw Hill, New York, 1964.

68. Sentance, W. "Occurrences of consecutive odd powerful numbers, *American Mathematics Monthly,* 88 (1981), 272-274.

69. Shanks, Daniel. *Solved and Unsolved Problems in Number Theory.* 1. Spartan Books, Washington, 1962.

70. Shenk, A. *Calculus and Analytic Geometry* (2nd ed.) Goodyear, Santa Monica, 1979.

71. Shoemaker, Richard. *Perfect Numbers*. National Council of Teachers of Mathematics, Reston, 1973.

72. Shultz, Harris. "An expected value problem, *Two-Year College Math. J.*, 10 (1979), 277-278.

73. Slowinski, David. "Searching for the 27th Mersenne prime," *J. Recreational Math.*, 11 (1978-79), 258-261.

74. Sobel, Robert. *IBM: Colossus in Transition*. Truman Tally Book, Times Books, New York, 1981.

75. Spiegel, M. "The beta function," *American Mathematics Monthly*, 58 (1951), 489-490.

76. Spiegel, Murray. *Probability and Statistics* (Schaums Outline Series), McGraw Hill, New York, 1975.

77. Venit, S. and R. Katz. "An analytic geometry approach to the least squares line of best fit," *Two-Year College Math. J.*, 11, (1980), 270-272.

78. Woo, W. "On generalized h-base," *The Two-Year College Math J.*, 6 (1975), 16.

79. Zagier, Don. "The first 50 million prime numbers," *The Mathematics Intelligencer,* 0 (1977), 7-19.

80. Zemgalis, Elmer. "Another proof of the arithmetic-geometric mean inequality," *Two-Year College Math. J.*, 10 (1979), 112-113.

81. Zill, Dennis. *A First Course in Differential Equations with Applications*. Prindle-Weber-Schmidt, Boston, 1979.

Index